THE GENE HUNTERS

THE
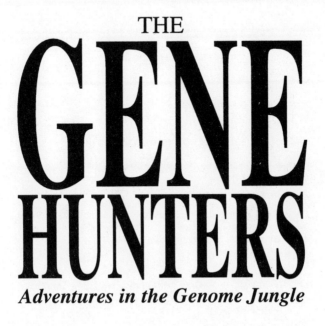
GENE
HUNTERS
Adventures in the Genome Jungle

WILLIAM COOKSON

AURUM PRESS

First published 1994 by Aurum Press Limited, 25 Bedford Avenue
London WC1B 3AT

A catalogue record of this book is available from the
British Library.

ISBN 1 85410 256 7

2 4 6 8 10 9 7 5 3 1
1995 1997 1998 1996 1994

Designed by Don Macpherson
Typeset by Computerset, Harmondsworth
Printed in Great Britain by Hartnolls Ltd, Bodmin

To the women in my life
Fiona, Caroline, Hannah and Alice
(in order of appearance)

CONTENTS

ACKNOWLEDGEMENTS

I make limited claims to originality in this book, and I hope only to pass on some of that which I have learned from so many others. I thank the many men and women who have conducted the science which I find so interesting, and those who have written about it in a manner to excite my attention. I apologize to those who feel I have neglected to acknowledge their contribution; it is a sin of omission which I will be pleased to correct if informed. In particular I acknowledge A. Sorsby, for his history of Gregor Mendel (*British Medical Journal* 1965; 1: 333–8); P. Froggatt and N.C. Nevin for their paper 'The "Law of Ancestral Heredity" and the Mendelian–Ancestrian controversy in England, 1889–1906' (*Journal of Medical Genetics* 1971; 8: 1–36); A.G. Cock for his paper 'William Bateson's rejection and eventual acceptance of chromosome theory' (*Annals of Science* 1983; 40: 19–59); *Science* magazine for reporting about the genetic map (Leslie Roberts, *Science* 1987; 238: 750–2), and the race for the cystic fibrosis gene (Leslie Roberts, *Science* 1988; 240: 141–4 and 282–5; Anne Gibbons for her article on sexual selection in primates (*Science* 1992; 255: 329–30); Alison Turnbul for her article 'Woman enough for the Games' (*New Scientist*, 15.9.1988); Leslie Roberts for the article 'Zeroing in on a breast cancer susceptibility gene (*Science* 1993; 259: 622–25); Jean Marx, (*Science* 1989; 245: 923–5); P.S. Harper and colleagues for their paper 'Anticipation in myotonic dystropy: new light on an old problem' (*American Journal of Human Genetics* 1992; 51: 10–16); M.R. Hayden for his monograph *Huntington's Chorea* (Springer Verlag, Berlin, 1981); William Poundstone for his book *Prisoner's Dilemma* (Oxford University Press, 1993); George Klein for his stories of the 48 chromosomes and the 'cuckoo's egg' of the oncogenes in his book *The Atheist and the Holy City*

(MIT Press, 1990); Robin McKie for the newspaper article 'Test for Alzheimer's disease may wreck lives' (*Observer* 1.1.94); D.W. Forrest for his biography, *Francis Galton: The Life and Work of a Victorian Genius* (Paul Elek, London, 1974); Robert Plomin for his paper 'The role of inheritance in behaviour' (*Science* 1990; 248: 183–88); John C. Greene for his book *Science, Ideology and the World View* (University of California Press, 1981); Jean Marx for his report 'Cell death studies yield cancer clues' (*Science* 1993; 259: 760–1); and Sir Peter Medwar for the enormous common sense of *Pluto's Republic* (Oxford University Press, 1984).

I would like to thank my agent, Sara Menguc, for encouraging me to write this book, and for guiding me as it grew. Finally I would like to thank my publisher, Piers Burnett, for his extremely helpful remarks on the typescript. The mistakes that remain are, unfortunately, my own responsibility.

'Curiouser and curiouser!'
cried Alice

PROLOGUE

Curiosity is a trait shared by many animals. Our own particular species, clever monkeys energetically prying into every nook and cranny of our planet and beyond, is consumed by curiosity. Our curiosity is narcissistic: we have always wanted to know why we are here, and what is our purpose in the grand order of things. Faced with mysteries beyond understanding, primitive man saw the action of gods: gods of lightning, gods of the river, gods of the sun. In these more sophisticated times we turn to the God of Science to understand ourselves. Now, as science begins to unpick the genetic code, we hope, or fear, that all our secrets are at last to be laid bare.

In this we will be disappointed. We are more than the sum of our genes. Science is not a god, only formalized curiosity. At heart, the real scientist is a child at play, picking things apart, shaking and touching and smelling them, learning what they are and how they work. When I was a little boy I too wanted to know how things functioned, paying the price more often than not for a disassembled tape recorder or radio. As a grown-up scientist I am still fascinated by the way things are put together, and so I have tried to show in the pages that follow the marvellous workings of the genome, the total of all our genes and our DNA. As well as the genes themselves, therefore, I hope also to show you about curiosity, and curiousness.

Our species takes great joy in giving things names and so I have indulged myself and have given the genetic things in this book their proper labels. I have included a short glossary at the end, should this practice threaten to defeat you.

Science cannot be understood without understanding the workings of scientists; character, both good and bad, will at times help the scientist and at times obstruct him. Where possible, therefore, I have tried to show the humanity, good and bad, of scientists themselves.

Much of this book concerns gene hunting, forays into the unknown reaches of the genome to discover the causes of human disease. In the first chapter I propound a theory of the origins of life. This is only a fable: the reason for its inclusion is to explain the basic structure and functions of DNA, RNA and proteins, and also to show you just how strange and beautiful the genome actually is. The chapters that follow describe the intellectual adventures of geneticists, who have over the past century made their alliances or warred over chromosomes, genetic maps and genes.

The early chapters concern the single gene disorders, such as muscular dystrophy, cystic fibrosis and Huntington's disease. There are two chapters on sex, as befits such an important subject. The first concerns the biological reasons for sex, and its evolution. This short chapter is more difficult than others, and could be skipped without loosing the thread of the book. However, the evolution of sex has been so wonderful and bizarre that I have felt obliged to include an account of it in my story. The subsequent chapter relates how the distinctions between men and women were traced back to a lilliputian gene on the Y chromosome.

Later come the complex genetic disorders, in which the actions of many genes interplay with the idiosyncratic effects of the outside world. Cancer, diabetes and schizophrenia are such illnesses. I have also written about asthma, my own research field, to illustrate the uncertainty of scientific progress.

There is much that is sensational in descriptions of genetics, both in the news media and in the books which cater to popular interest. This sensationalism reflects the considerable expectations that have followed from our ability to dismantle and to manipulate our genes. Unfortunately it distorts the reality, which is that advances in the treatment and prevention of genetic disease may follow only slowly and painfully after the 'breakthrough' of isolating a disease gene. In the later parts of the book, therefore, I have tried to explain the many difficulties that have to be overcome before genetics improves the lot of the sick.

The obverse of better medicine is that genetic knowledge may lead to actual harm. Genetic screening, for disease genes, and for desirable traits such as male- or female-ness or intelligence, raises many complex ethical issues. In my last three chapters I have tried to discuss the problems, usually without presuming to judge what is right and what is wrong. It is the public who must in the end decide how genetic knowledge may be utilized. My part is only to give as much information as I can.

CHAPTER 1

DNA

To begin at the very beginning, let us assume that life started in a pool. The pool was near the equator. During the day and in the sun it was very hot, nearly boiling because of the thinness of the atmosphere. At night it was cooler. In the pool, and many others like it, the elements abundant in the earth's crust had dissolved into the water. Atoms of carbon, oxygen, hydrogen and nitrogen had combined to make simple molecules, such as acids, sugars and alcohols.

In the pools, some molecules stuck to each other, making new and more involved structures. Every time the sun shone and the pool heated up, the molecules tended to separate, only to reform in new combinations when the temperature fell in the evening. Stable molecules broke up less often than their unstable companions, and so became more common in the soup. As time passed, the soup in the pool matured and quite complicated molecules came into being. Eventually, nevertheless, some state of equilibrium between all possible compounds was reached, and the process could move no further.

However, some pools may have been rather special. They contained in relative abundance four or five quite similar molecules. The molecules were rings of carbon and nitrogen, with bits sticking out at the sides that gave them stability. These molecules were identical to those found in the nuclei of living cells, which is why they now are called 'nucleotides'. Their individual names are thymine, adenine, cytosine and guanine. It is easier to call them 'T', 'A', 'C' and 'G'.

The acidity and the concentration of salt in these pools would have encouraged the nucleotides to form stable compounds with a sugar called deoxyribose. A bit of phosphate in the solution meant the deoxyribose molecules were attracted to each other. As a result they tended to form long chains. The deoxyribose formed the backbone of the chain, with the nucleotides, T, A, C and G, along the edge, like legs on a millipede. As the pool cooled the chains grew, and because they were stable would resist breaking up again next day. Some breaking up would, however, occur, and the order of the nucleotides on the chain could differ from evening to evening.

Extraordinary events then followed. This was because not only did the sugars stick to each other, but so too did the nucleotides. C stuck to G and A to T. Furthermore, because of a fortunate matching of structures, C stuck only to G, and A exclusively to T. The tendency to nucleotide adhesion meant that chains tended to form in pairs, with the sugar chains running along the edges, and the matching nucleotides meeting in the middle like the teeth in a zipper. This is the structure now recognized as DNA.

So what is the importance of this zipper structure? It could not be thought of as alive, except for the unending cycle of the heat of the sun and the cool of the night. When the pool became hot each morning, the strange, long, double-stranded molecule would begin to break up. However, because the links in the deoxyribose backbone were tougher than those between the nucleotide teeth, the molecule would not shatter into pieces, but would instead unzip into two separate chains which would then drift apart. With the coming of the cooling night, stray nucleotide molecules would stick to the two original chains, matching C to A and G to T, and gradually building up two more complete chains. In the morning there would be four chains in two pairs; after the next day and night, eight chains; after the next night sixteen. At the end of a month there would be 1,000,000,000 identical chains and their matching complements. It is at this point that life had begun, because for the first time a molecule had the capacity to replicate itself infinitely.

In the various pools there were different strings of nucleic acids, each with its own particular order of As and Cs and Ts and Gs. The order of the nucleic acids on a DNA chain is known as the sequence. The matching order on the other strand is called the complimentary sequence. In the sequence of nucleic acids was eventually to be written all of the instructions for life on earth.

In most pools there would be billions of different strings, each busily copying itself. At some stage certain strings must have emerged as more stable than others, and thus able to make copies of themselves more efficiently than the others. Eventually, as all the sugars and the nucleic acids in the soup were made up into chains, the more stable molecules would persist at the expense of their less stable companions. The successful DNA molecules might have been circles, or folded back on themselves like hairpins. However, even within the more stable forms of DNA, changes would be occurring all the time, because copying would be imperfect, or because other molecules interfered with the DNA at a particular point.

The occurrence of small changes meant that the structure of the DNA could be improved without loosing any advantages that might already have been accrued. The improvement might involve the formation of an even more stable structure, but it might also mean that the molecule could copy more easily or faithfully than other DNA conformations.

If we think of the DNA as being alive at this point, we can recognize it to be a highly ambitious molecule, replicating itself left, right and centre at the expense of other molecules. It was, however, limited in its ambition because all it could really do was divide in two and copy itself. The chance for DNA to extend its influence came about because there were other molecules in the pool, particularly those that shared a tendency to form long chains. These molecules were RNA and amino acids.

Like DNA, RNA has a sugar backbone and nucleotides along the edge; but in RNA the sugar is slightly different. The RNA sugar is a ribose, which contains one more hydrogen and one more oxygen molecule than the DNA sugar, deoxyribose. RNA also differs from DNA because it uses the nucleotide base uracil ('U') instead of T. Because U binds to A in the same way that T does, RNA can form a matching strand to DNA, or two matching RNA strands; but quite often in advanced life forms we find RNA strands on their own without a partner. Single-stranded RNA is particularly different from DNA in one important aspect: it can fold into more complex shapes, such as four-leaf clovers, or blobs, and because of this it can act like an enzyme. Enzymes are molecules that can do things: it is, for example, enzymes in your washing powder which dissolve the stains. In the pool, some RNA molecules could cut and rearrange other molecules, or even stretches of their own structure.

We recognize today that many types of RNA still possess this enzymatic quality, and can cut and splice themselves in a highly specific way. For this reason it has been proposed that the first molecules of life were not DNA but RNA. Personally I doubt that RNA was the first molecule of life because it is so fragile. You only have to look sideways at RNA and, heartbreakingly, it disappears, as any researcher who works with it will tell you. DNA, on the other hand, is as tough as old boots. Molecular archaeologists are able to extract and recognize DNA from bones that are thousands of years old, and from plants as old as ten thousand years. DNA can even be found, albeit in a very degraded form, from insects who were caught in the sap of trees thirty million years ago, or from fossilized Jurassic fish. Consequently my money is on the DNA. To have survived and prospered in a boiling pool, life could only have been based on something as robust as DNA.

So there in the pool was the tough old DNA, which at the end of the hot day was in single strands. During the cool night the delicate RNA would form along the DNA, copying it as it went. When it was hot the next morning the RNA would separate from the DNA, free to act as an enzyme and do clever things. The RNA would eventually fall to pieces if it was separated from the DNA, but the DNA persisted and could make an identical RNA molecule the next day. Thus the DNA had gained control of the RNA: RNA was the first slave of DNA.

Although DNA and RNA together had the potential to develop more and more complex RNA structures, they were still in the pool. In order to escape, a third partner was needed. DNA learned how to get out of the pool when it came into the company of amino acids.

Amino acids are other simple molecules, made up of the same basic atoms as nucleotides and sugars. Like DNA and RNA, amino acids can form chains and, although many widely differing varieties exist, they all possess the capacity to bind to other amino acids using similar joints, rather as the differently shaped and coloured blocks in a Lego set share a common fastening system. Chains of amino acids are bulkier than DNA or RNA and, because of the variety of amino acids, there are potentially thousands of ways of folding and twisting these chains into different shapes. This huge variety of chains has an equal number of possible chemical and physical properties.

Chains of amino acids are called proteins. We are made of proteins: skin is protein, muscles are protein, bones are protein with calcium. Some proteins can open and close, or twist and untwist, or lengthen or shorten according to which other molecules they are touching. Your heart is beating in your chest because of proteins, and the electrical currents with which you are thinking as you read this paragraph and prepare to turn the page are controlled by proteins. In fact, nearly all the myriad functions that keep your cells fat and healthy are carried out by proteins.

The RNA in the pool, because of its ability to form molecules of different shapes and sizes, was able to react with the amino acids. What was important was that particular, short RNA sequences preferential-ly stuck to specific amino acids. Not all amino acids could be picked up by RNA sequences, which is why living things today use only a few amino acids, about twenty. Thus a DNA sequence would consistently result in RNA that would consistently bind to the same amino acid. The DNA was thus in command of the amino acids: whenever a given sequence appeared in the DNA a particular amino acid would subse-quently be plucked out of the pool and held by an RNA molecule. The RNA molecules that hold the amino acids are now known as transfer RNA, or t-RNA.

Transfer RNAs are made from a single chain of RNA. Some of the sequences in the chain match each other, so the RNA is folded into a cross. The four arms of the cross are made up of two complimentary strands, but at the top and sides of the cross the RNA forms open loops, like three leaves on a clover. At the bottom of the cross (or the stem of the clover), the amino acid binding site hangs free. At the head of the t-RNA is a loop of seven nucleic acids, the middle three of which are exposed outside the molecule. In the primeval t-RNA the open loop had a vital consequence: it could bind to the DNA or RNA that had the complimentary or matching three nucleotide sequence.

The t-RNA thus had two regions that could react with other mole-cules. The bottom of the cross could pick up an amino acid. The open loop at the top could bind to a particular sequence of three nucleotides. This sequence of three nucleotides would then, through the medium of t-RNA, be associated with a particular amino acid. Through the t-RNA the DNA could reach out to the amino acids, the building blocks of protein.

The relationship between a triplet of DNA nucleotides and an amino acid is the basis of the genetic code. The four nucleotides C, A, T and G can be arranged into 64 possible three-letter 'words', such as CCC, CAC, CTC, CGC, CAT and so on. The three-letter words are known as 'codons'. A total of 64 codons were more than enough to match with the twenty amino acids that could be picked up by the tail of the t-RNA. The DNA sequence that made the first t-RNA can be considered the first gene, the first sophisticated use of sequence infor-mation to make an object acting directly upon the environment.

Once DNA could make a t-RNA that would grasp amino acids, another set of sequences could cause the t-RNAs and their individual amino acids to line up in order. t-RNAs do not line up on DNA, but instead line up on a string of RNA, called messenger RNA or m-RNA.

m-RNA in a mixture with t-RNAs and amino acids could cause the amino acids to arrange themselves in the same order every time the m-RNA was copied from the DNA. The DNA then had the crude apparatus to make a protein, as a protein is simply a string of amino acids. As there are 64 triplet codons and only twenty amino acids are used, there is a redundancy in the genetic code, and many codons do not correspond to amino acids. Some of these are simple nonsense, but others mark the beginning and ends of proteins ('start' and 'stop' codons).

That 'DNA makes RNA makes protein' has been called the 'Central Paradigm of Genetics'. Anything with so grand a name cannot be entirely correct. The Central Paradigm has its faults, which we will hear about later, but for most of the time it holds true.

To return to the pool, once DNA had control of protein there was no stopping it. With protein it could wrap itself in a membrane to keep out unwanted chemicals, thus forming the simplest of living cells. In the cell it could make enzymes that helped it to replicate itself away from the heat of the sun. It could make proteins that helped to feed energy to the other proteins, and proteins to move the cell around, and proteins to join different cells together, until, to cut a long story short, it ended with life as we know it today.

Because we are clever, and we are built by DNA, we like to think of the DNA as being as clever as we are. This is wrong. DNA is not at all clever, it is just tough. The sequences that do mean something have only come together by chance. In the primeval pool, simple chemistry would have led to the formation of the DNA and the RNA. A pool, or thousands of pools with billions of random DNA and RNA sequences, would have given rise reasonably often to t-RNA-like molecules. Interactions with amino acids and simple proteins are also likely to have been common. It would have taken much longer for the useful proteins to appear. For this to happen luck was necessary; just as the monks in a monastery might take many, many meals around their communal table before, one day, they all sat round the table in such an order that their first initials spelt the name of the winner of the 4.40 race at Doncaster. This sort of random event obviously does not occur very often, and it might be thousands of years before another order of monks would spell out the name of another horse. The sequence of nucleic acids that spells the code for a protein is usually considered the most basic gene. Strictly speaking, however, the first genes made RNA rather than protein.

Only over millions of years would chance create enough words from monks' initials to make a proper racing program. Similarly in the pool, only after millions of years would there be enough sense in the DNA to make a living cell. A cell is characterized by a membrane, like a cellophane wrapper separating the DNA and RNA and their proteins from the outside environment. The early cells possessed several basic functions. The first of these, the process by which DNA copies itself, was called replication, and was mediated by enzymes which made the process much more efficient, and less prone to mistakes in copying.

The DNA in the primitive cell was not contained in multiple separate strands, each carrying a gene or two, but organized into one strand containing all the genes. This meant that replication of all the genes could be carried out simultaneously. To avoid damage, the long strand of DNA was condensed into a compact bundle. These bundles of DNA are known as chromosomes. In the first cell there was only one chromosome, but higher organisms such as animals usually have several. Chromosomes can be seen under the microscope. The total of the DNA in a cell is known as its genome. In a cell with only one chromosome the genome and the chromosome are the same thing. If there are many chromosomes then the genome corresponds to all of them. The genome of a many-celled organism such as a plant or an animal is replicated in each of its cells. It is therefore possible to refer to 'the mouse genome' or 'the human genome'.

The early cells would also have genes for other functions: controlling cell division and metabolism, the extraction of energy from the environment to fuel the cellular processes. Bacteria are the closest living relatives of the early cells, and still form the most abundant form of life on earth. Billions of years later, we still share the same fundamental functions for metabolism and DNA replication as bacteria.

We are usually taught in school that higher forms of life have evolved from these simple cells, and that the process of evolution is carried out by mutation. A gene uses RNA to make a protein. Every protein has its own gene. Changes in the DNA sequence of the gene are known as mutations. If a mutation alters a particular triplet codon, then there will be a different amino acid in the protein for which the gene codes. Altering the amino acid changes the protein, so genetic mutations can alter the function of a protein. If the protein makes your hair red, then a mutation might make your hair blond. If the protein is part of muscle, then the mutation can give you muscular dystrophy.

Usually mutations are harmful, and if they are very harmful the DNA carrying them is soon lost, because the organism it controls dies.

Very occasionally a mutation is helpful, so that it makes a better protein, for better muscles or blue eyes, for example. Because mutation is so random, you might imagine that to evolve from scratch all the genes used by different living things would take an infinite length of time. The time taken has not been infinite because DNA developed a few short cuts.

The most useful short cut is called duplication. Duplication simply means that when DNA is copying itself it sometimes makes an enormous mistake: two copies of a gene are made, one after the other on the same strand of new DNA. Thereafter the DNA will always have two genes for the same protein. It could in the future make twice as much protein. However, the second gene would also be free to mutate into something different and worthwhile, while the first gene carried on with its usual task.

As more genes are discovered it becomes more and more obvious how much duplication there is in our DNA. Whole families of genes are often found, all bearing a resemblance to each other. They have all evolved from a single common ancestor, though they may now have completely different actions. Sometimes the members of gene families are still together on a chromosome like Italians. At other times they are Irish and spread in all directions to the far corners of the genome.

Whole chromosomes have also duplicated themselves. In our species the current number of chromosomes has come about by a series of duplications or quadruplications with the odd loss here and there. Some human chromosomes are still very like others in their structure and the type and order of the genes found on them. Any variation in the number of chromosomes in higher mammals is usually lethal, or nearly lethal. By contrast, plant chromosomes are much less restrained, and two or four or eight copies of chromosomes are quite common. Plant chromosomes are also completely promiscuous, happily mixing and matching away between species to form hybrids with entirely new shapes or properties.

Some duplicated genes lose their function altogether. These 'pseudogenes' are ruins of genes. They still look like genes, and their sequence mostly makes sense, but they may be interrupted with nonsense mutations, or stop in strange places. Pseudogenes may be viewed as nature's mistakes. Alternatively it may be in these genes that evolution is taking place: DNA is experimenting at making new proteins without throwing away the old ones.

As well as duplication, DNA has another trick up its sleeve: conservation. Conservation means that the genes for the most basic functions of life are very similar in most living things. Even in the most advanced forms of life the conservation persists. Genes that supply energy, or control the synthesis of DNA and proteins are similar in bacteria and cows. The genes that control the earliest events in fruit fly development are recognizable in human beings. In the higher mammals the conservation is profound: we have 90% of the same genes as a mouse, and 99% of the same genes as a chimpanzee. This similarity is carried through to the chromosomes. The same genes group in the same order on similar chromosomes in many mammals. Our chromosomes, for example, are 78% the same as cats, and 82% the same as sheep. On the other hand, we are only 38% the same as mice, and so mouse chromosomes look very different from ours. Nevertheless, only forty cuts and pastes could reassemble the mouse genes to be in the same order on their chromosomes as on our own.

As knowledge has advanced in the last decade, the structure of DNA is being shown to be stranger and still stranger. In medical school I learned to think of genes as beads on a string. One bead, one gene, one protein, all very simple and straightforward. Unfortunately, DNA is made of two strands which, if complimentary, are not identical. There is, therefore, a possibility that one gene might run in one direction on one strand, while another gene runs in the opposite direction on the second strand. In fact this extremely confusing process does take place.

The occurrence of small genes within big genes complicates matters further. So, too, do genes that are one length at one time, and another length at another. Such genes make more than one protein, depending on the tissue (such as muscle or brain) in which they are acting, and potentially one protein might be completely normal and the other abnormal.

Obviously, as DNA evolved more complex proteins, it became increasingly important to turn genes on and off. In any higher organism, the control of genes is a matter of staggering complexity. The 100,000 human genes are not all turned on at once. Many of them are to do with development, the growth of an infant from a solitary fertilized egg. They may only be turned on for a few hours, right at the beginning of life, and then lie dormant indefinitely. Turning them on at the wrong time can cause cell death, or uncontrolled cell growth and cancer.

The reading of a gene to make messenger RNA is called transcription. At the right time an enzyme, starting at the origin of the gene, moves along the DNA, copying it to m-RNA as it goes. At the beginning of the gene, often a little distance 'upstream', are sequences of DNA called 'promoters'. Some promoters are downstream. It's not known how promoters work, but without them many normal genes will not make m-RNA, and so will not make proteins, meaning they are as good as useless.

Genes are likely to control each other through their promoters. When a gene is turned off, a protein sits on top of the promoter like a policeman on a demonstrator. Only when the policeman is gone can the demonstrator express himself, and can the transcription enzyme get to the gene to make m-RNA.

To make a protein, the order of the letters in the DNA code has to make sense. A 'nonsense mutation', a bit of code that does not correspond to an amino acid, in the middle of the gene prevents it from being transcribed into a normal protein. It is like an error in a computer program. At the end of the gene is a 'stop codon', a three-letter word that tells the enzyme to stop copying the DNA to RNA. A mutation can put a false stop message in the middle of a gene, so that only half a protein is made.

Stretches of DNA that make sense, and will call up a protein, are called 'open reading frames'. Before 1977 it was believed that any given gene consisted of a single open reading frame, and that this reading frame made sense from beginning to end. This is actually the case in bacteria. However, when working with viruses that infected humans it was found that m-RNA did not always match the DNA it was copied from. The difference was traced to areas of nonsense within the open reading frame of the gene. The areas of nonsense are known now as 'introns', and the areas that make sense known as 'exons'. The discoverers of introns, Richard Roberts and Philip Sharp, received the Nobel Prize in 1993.

This arrangement is like introducing a few thousand lines of gibberish here and there into a computer program. At first sight it seems unlikely that such a gene could effectively code for a protein. However, the whole gene, rubbish and all, is copied from the DNA to m-RNA. In the RNA the nonsense introns are excised by an RNA enzyme, and the remaining exons spliced together to leave a single open reading frame and a whole gene.

The mixtures of introns and exons are ubiquitous in organisms other than bacteria, so it is safe to assume they must serve some purpose. It is possible that exons represent the earliest small proteins. These have been christened 'ancient conserved regions' (ACRs) by P. Green. These ACRs are hypothesized to have served as building

blocks for more complex proteins. The exon–intron structure of genes
aided evolution, by allowing the shuffling of different ACR exons
around to make novel proteins. An analysis of the structure of proteins
from a wide variety of life forms (bacteria, plants, fungi, slime moulds,
vertebrates, insects and worms) was carried out by J.-M. Claverie in
Bethesda, Maryland. He found that there were only about 550 basic
protein motifs, from which most other proteins could be assembled.
Most were about eighty amino acids long. This number of ACRs is
quite small enough to have explained the current state of evolution
within the known habitable span of our planet.

Exon shuffling does not only take place in evolution. In white blood
cells the genes that make antibodies are called 'jumping genes'.
Jumping genes use different exons that are deliberately mixed and
matched to make possible the huge variety of antibodies needed to
deal with any potential infection.

With the exception of the genes, a great deal of our DNA seems to
be nonsense. It is not known whether the nonsense serves some arcane
purpose, or whether it is detritus. Even among the detritus, however,
are many repeating patterns: the genome is everywhere full of copies.
The two most common types of copies are called LINES and SINES,
and are very odd indeed.

LINES are 'long interspersed sequences'. About 4% of all mammalian DNA is made up of LINES. Extraordinarily, the structure of
LINES is related to a class of viruses known as retroviruses.
Retroviruses are anarchists which break the Central Paradigm that
DNA makes RNA makes protein. They are particularly nasty because
they hijack the machinery of DNA and make it work backwards. They
only have three or four genes, one of which makes an enzyme called
'reverse transcriptase'. Reverse transcriptase copies the virus RNA
back into human DNA. The DNA then works away making virus proteins and new viruses. Once reverse transcription has happened the
virus sequence is part of the DNA of the infected cell. All the progeny
of the affected cell will henceforth carry the corrupt DNA.

Mercifully, what usually happens is that the infected cell dies, perhaps as the result of immunity mechanisms, and the virus DNA dies
with it. However, if the reverse transcription has taken place into a
sperm or ovum (known as a germ-line cell) then the virus DNA
becomes part of every cell in a child fertilized by the sperm. The virus
genes are then passed to the child's children and grandchildren and so
on for ever.

LINES are like very primitive retroviruses. Although they do not make protein, they can copy themselves into other parts of the genome, occasionally disrupting normal genes and causing disease. There are between twenty and fifty thousand LINE sequences in our genomes. Although they do not make anything, the replication of these DNA segments makes them behave like 'selfish genes': their only concern is to look after themselves. We shall return to the subject of selfish genes later.

SINES (short interspersed sequences) are another kind of selfish genes that are spread through the genome. It is not known why they are there, but they have been successful enough to make up 3% of our genome. Many of them are in families, and look as though they were once part of a very common gene that makes an RNA enzyme for transcription. SINES and LINES do not confuse gene hunters by pretending to be genes but they can cause great technical difficulties when, as often happens, a researcher is trying to find a match for a piece of DNA somewhere in the genome. The presence of a repeat in his piece of DNA simply means that it matches everywhere.

LINES may be related to pieces of DNA which move about the genome without ever leaving the nucleus and which are known as transposable genetic elements, or transposons. Their existence was first guessed at by a plant geneticist called Barbara McClintock. She was born in 1902, and made her considerable early scientific reputation in the study of plant chromosomes under the microscope. By 1945 she was President of the Genetics Society of America, universally respected for the quality of her work and loved for her qualities as a teacher.

In 1942 she went to the Cold Spring Harbor Laboratory in New York. Maize was of particular interest to McClintock and other plant geneticists. The Indians who fed on the maize liked the many colours that showed in the kernels, and had maintained their variety in the crops that they cultured. The colours and other traits could be used to track the behaviour of the maize genes. Because she was a meticulous observer McClintock could discern patterns in the pigmentation that showed particular genes were being switched on or off in an irregular way. These results she correctly interpreted as due to 'controlling elements' that could move around the maize genome.

This was before anyone knew what a gene was, or even that DNA was the genetic material. McClintock's concept of a dynamic genome was therefore incomprehensible to almost every other geneticist. Her findings were ignored or, worse still, treated with derision. This reaction to her work became so painful that she refused to submit any papers for publication. Her most important experiments were described only in the house journal of her institution, with a reader-ship limited to a handful of close colleagues. In the mid-1960s transposable elements were found in the DNA of bacteria, by the 1970s in the DNA of plants, and by the 1980s in mammals. McClintock was awarded the Nobel prize in 1983; she was then 81, and still scientifi-cally active and productive. In 1993 a transposable element was found to have caused Von Recklinghausen's disease, a human genetic illness affecting the skin and nerves.

Barbara McClintock was and is an icon for young scientists: she was an extraordinary observer of facts that others dismissed as artefacts or as insignificant. She could invest these apparently trivial observations with their true significance. When her love for her subject compelled her to carry on her work even in the face of severe opposition, she became the first scientist to glimpse just how extraordinary is the structure of the genome.

The term 'the blind watchmaker' has been used to describe the role of evolution in constructing genes. We can better understand the nature of the genome when we consider how a blind man might fare at the assembling of watches. In reality the genome does not resemble a neatly ticking timepiece, but rather a surreal junkyard: all the pieces of the watch are in the yard somewhere, but the second hand is not necessarily anywhere near either the minute hand or the mainspring, and the watchmaker has forgotten that he has made six winders, only two of which still work. Nevertheless, the hundred thousand moving parts of the watch run together completely smoothly, and the watch keeps perfect time.

The genome is a shambles because, unlike the works of Shakespeare, the book of life was constructed by a series of chance events, the equivalent of millions of monkeys hammering away at their word-processors to produce a *Hamlet*. Only one-tenth of the book of life actually says anything intelligible. The chapters are in no par-ticular order, and even the paragraphs are interrupted by stretches of gibberish. The book reads both forward and backwards, and the same length of code can spell out different genes in one or both directions.

Thus the genome is a rather exotic place. It is through this prepos-terous terrain that the gene hunter must struggle as he tracks his prey.

The blind watchmaker

CHAPTER 2

GENE HUNTING

Within 3 billion letters there are over 100,000 genes

— HGP

Already complete

Four thousand million years of evolution have moved DNA from the pool to our present selves. The genes contained in our DNA have given us the curiosity and intelligence to unravel how they control us. The human genome, the sum of all our DNA, will be sequenced by the year 2010 when the international programme known as the Human Genome Project is completed. Sequencing means that the three billion letters of our genetic code will have been read from beginning to end. Within the three billion letters will be found more than a hundred thousand genes.

It is a mistake to imagine that the sequencing of the genome will result in immediate comprehension of how the genome works: it will not. Sequencing the genome is a technical *tour de force*, but not much more than that. At the end of the exercise the code for all our genes will be available, but this information is not nearly enough. It will be rather like having all the parts of a jet fighter, down to the last nut, bolt and rivet, laid out for inspection. It is quite easy to list the parts, and arrange them according to size or to shape, or as to whether they have holes in them or not. Understanding what the parts do, or how they are put together to make a working machine, is an altogether different matter. For this reason it will take forty or fifty years after the end of the genome project for functions to be ascribed to most genes.

For the sick, however, fifty years is too long. There are about three and a half thousand known genetic illnesses, due to mutations that alter or destroy the function of genes. Until relatively recently genetic illness was of little more than academic interest. Genetic diseases were perceived as obscure maladies producing bizarre and deformed individuals. The committed scientist might succeed in naming a syndrome for himself, but the only therapeutic possibilities were the counselling *'reverse genetics or position cloning'*

CLONING

of parents as to the risk of bearing a second affected child.

The invention of cloning, the manipulation of genes and the genetic material at will in a test tube, took place in 1973. The ability to sequence genes, to read their genetic code, arose at about the same time. Cloning and sequencing completely changed the way that scientists could examine disease. The new genetics has followed from the technology of cloning genes and means that it is now possible, in theory at least, to find the mistake causing any genetic disease, by a process known as 'reverse genetics' or 'positional cloning'.

Reverse genetics is gene hunting, going out into the genome and getting your gene. Of the hundred thousand human genes, only five hundred have been captured and cloned. By far the greater part of the realm of DNA is an unexplored world. The world holds continents. In the continents are jungles, exotic ruins, and wild and savage beasts. A lucky few who enter the genome might find treasure, fame and limitless wealth. Others will find mystery, danger, despair and humiliation. Some explorers will set out with well-ordered expeditions, their wagons loaded with supplies and weapons. Others will amble off into the wilderness with only a rifle. Success will depend on luck as much as on good provisions.

Gene hunters have already discovered the genes for the most common of the 'simple' genetic disorders. The greatest successes have been the discovery of the genes for muscular dystrophy and cystic fibrosis. Cystic fibrosis destroys the lungs and pancreas of one child in four hundred; muscular dystrophy cripples one boy in a thousand. Both result in the premature death of children who are otherwise completely normal. Until their genes were discovered, the origin of these illnesses was quite unknown. Now, although in honesty it must be acknowledged that cures are not imminent, therapy can for the first time be aimed at a known cause.

There are many more single gene disorders, the most common of which have been found. The most important diseases remaining are those which are likely to affect many of us at some time in our lives. Illnesses such as cancer, diabetes, hypertension, asthma and schizophrenia are known to be due partly to defects or variants in genes. To what extent each disease is due to genes, or even how many genes are at fault, is not known. Because they are difficult to understand, these illnesses are lumped together under the label of 'complex'. Complex in this context is an understated way of saying 'too hard to explain'. Nevertheless, the prevalence of these illnesses means that the greatest rewards of genetic research may result from their understanding.

Gene hunting is first of all a serious science. The successful gene hunts have all been conducted by big groups of scientists in furious competition with each other. It is very expensive research. The reagents used to carry out experiments, the ingredients for the exotic cookery that is molecular biology, are terribly costly; equipment and salaries add much, much more to the bill. Nor does the expense stop there: over a thousand families were tested in the search for the cystic fibrosis gene. That means more than four thousand people had to be interviewed and rigorously examined for presence of the disease before any study could be made of their DNA in the laboratory. For more complex illnesses ten times that number might need to found. Nevertheless, the belief that genetics may answer previously insoluble medical problems has meant that genetic research is generously funded, even as other areas of research endure cuts in support.

A gene hunt begins with the search for genetic linkage, proof that a disease gene belongs to part of a particular chromosome. In most cells of the human body there are 46 chromosomes. These are made up of 23 pairs. Although they vary in size, they contain on average 4,000 genes each, and 120 million letters of the genetic code. If a gene is abnormal and causes disease because one letter of its code is misspelt (as is commonly the case), then finding that one wrong spelling is comparable to finding one individual amongst all the people of the world.

Establishing genetic linkage is the equivalent of discovering that the individual is in England or California: one of 50,000,000 instead of one of 3,000,000,000. Assuming that you are an alien from Alpha Centauri, and you have arrived on earth to talk to a specific person who has got your sister into trouble, you will need help to find him. Maps and telephone directories will be your most important tools.

CHAPTER 3

THE MAP

To understand the human genetic map, picture yourself dropped by parachute into a jungle about which you know nothing. How would you go about finding your way around? You would probably climb to the nearest high place and look around. Maps begin with recognizable landmarks, places that are references in the wilderness, from which explorers can spread out and to which they can find their way back. In the world of genes, the first landmarks were the chromosomes. A chromosome is a packet of DNA, a long string of genes that can actually be seen under the microscope. Different species of animals and plants have different numbers of chromosomes. In humans there are 22 chromosome pairs, besides the X and Y sex chromosomes which make up the twenty third pair.

In 1973, which is not all that long ago, a group of researchers met at Yale in New Haven on the Connecticut coast. Their purpose was to summarize all that was known of the human genetic map. The table they drew up ('DATA1') contained the genes then known to lie in particular positions on chromosomes. Because it was not yet possible to isolate and sequence the DNA that coded for genes, the presence of genes had to be inferred by their effects. Thus the table contained 27 'Mendelian markers'. These were mainly known associations between specific diseases and particular chromosomes. Also on the table were 57 *'in vitro* markers', chromosomal locations, associated with the production of the proteins in our blood, where, it was deduced from small differences in the protein products, DNA sequences must differ between one individual and another. At the time, it was only an assumption that the diseases and the differences in the proteins were due to genetic variation on particular chromosomes, although the assumptions were later to be proved correct. A mere 84 points are not many references with which to map a territory that contains a hundred thousand genes. It was not possible to tell chromosomes 4 and 5 apart. There were no markers at all on chromosomes 3, 8 or 9. The researchers called their meeting 'Human Gene Mapping 1' and resolved to meet again at regular intervals to review the progress.

In 1989, as a tyro geneticist, I attended the tenth gene mapping conference ('HGM10'). HGM10, like HGM1, was held in New Haven. New Haven is a university town, an Oxbridge transmogrified by transposition to Connecticut, with the difference that a hundred yards away from the university campus things rapidly become very seedy, and two hundred yards from campus they are downright alarming. The streets were full of construction workers in hard hats. At night the sounds of heavy machinery mingled with the roar of high-powered pickup trucks and the scream of police sirens, just like in the movies. Most of the delegates stayed in the student accommodation. I was not sure whether to be alarmed or reassured by the guns slung at the waists of the campus custodians. Perhaps at Oxford the students might sleep better for the knowledge that old Ned at the college gate was tooled up, but I suspect not.

About 700 delegates attended the meeting, the Europeans among them totally astounded by the generosity and depth of American hospitality. I do not think I have ever seen food in such quantity. I fell in with a bad crowd from St Mary's in London, discovering over several evenings that Budweiser beer had rather more bite than its taste suggested. During the day there were lectures by the great and good of gene mapping, or open meetings of the chromosome committees.

The HGM meeting was so big that even Jim was there. Jim Watson had discovered how DNA encoded the secret of life. We were all in New Haven because of what Jim and Francis Crick had done. At a clambake on the beach on the third evening of the conference Jim was wandering round, the prototypical eccentric scientist, hair and eyes all over the place. Wherever he walked, people would grab him and beg for a photograph to be taken with them. He never said no, smiling benignly and shaking hands as the flashes popped and his image was preserved for pride of place in another stranger's family album. He may even have enjoyed the experience. I am not sure. He certainly enjoyed the experience more than the very fresh lobsters that were trying to crawl off the super-heated barbecue coals. I didn't eat anything that night, but I went to bed happy because Jim had brushed past and his coat had touched me.

Most of the real business of the conference took place behind closed doors in committees, where information was sifted and collated to make up a map that could be published. This list of mapped markers would serve as the reference for anyone who wanted to map genes to a given region during the next two years.

By HGM10 the genetic map had become so dense, and the input of new information so vast, that the maps of the different chromosomes had been divided up between committees, one committee per chromosome. The committees spent most of their time wrestling with the data and with the computers set up to feed information into a central database. By HGM11, held two years later in London, the volume of information had outgrown the meeting. It was therefore decided that in future the individual chromosome committees would have to hold their own international meetings.

At HGM10 the final map contained 4362 segments of DNA which had been assigned to particular chromosomes. Of these, 1886 were suitable for use in mapping. Six years before, at HGM7, there were 319 such segments, of which only 130 were suitable for mapping. To do the sums, in the first ten years of HGM meetings 231 new points were found on the map; in the next six years the number was 4043. The genetic map had taken off.

Although the idea of a map now seems obvious, it had a humble beginning. Even the importance of the chromosomes was slow to be recognized. For most of the time, because they are so extremely long and thin, chromosomes are invisible, and lie in the nucleus of the cell like a mass of very fine spaghetti. However, just before the cell divides, the chromosomes shorten dramatically to form thick stick-like structures. These condensed chromosomes are easily visible with a microscope. They are most easily seen in tissues that are growing rapidly, such as the roots of plants, or the lining of the intestine. By 1855 the structural consistency of chromosomes through generations was established; by the early 1870s their numerical constancy in species was discovered. However, although these facts were agreed, no-one had any idea as to the chromosomes' function, and their presence was no more than an interesting phenomenon. Their significance became clearer soon after 1900. It was in this year that the great 'rediscovery' of Mendel took place.

Johan Mendel laid the foundations of genetics, and so we should know something about him; he was the second of three children, born in 1822 into a peasant family from Moravia, which was then part of Austria, and is now part of the Czech Republic. At 21 he entered holy orders, taking the monastic name of Gregor. He did not prosper as a parish priest, and was pushed into teaching. He was, however, unable to pass university examinations in Vienna, and remained an assistant teacher until the age of 44. His late development should be a lesson to those who use the name of genetics to justify pernicious assumptions about innate ability.

At the age of 34 Mendel began his experiments on breeding edible peas. By the mid-nineteenth century, plant breeders were accustomed to carrying out crosses or hybridizations between different plants. However, these were of the 'try it and see' variety, with no scientific basis or underlying theory to predict their outcome. The first thing that Mendel saw was that crosses between two varieties of plant produced the same types of hybrid offspring with 'striking regularity'. A cross between the same two varieties of plant would always produce the same sorts of offspring. This might seem a trivial observation, but by making it Mendel had seen that heredity operated in a constant and measurable way. No-one before Mendel had had this insight. It was to be fifty years before anyone else would even approximate this brilliant inspiration.

Mendel experimented by making crosses between 34 varieties of peas. The

> ...peas selected for crossing showed differences in the length and colour of the stem; in the size and form of the leaves; in the position, colour, and size of the flowers; in the colour, form and size of the pods; in the form and size of the seeds; and in the colour of the seed coats and the albumen.

Mendel observed that some of these 'characters' did not show a '...sharp and certain separation, since the difference is of a "more or less" nature, which is often difficult to define.' For his experiments he chose only '...characters which stand out clearly and definitely in the plants'.

The decision to study only 'clear and definite' characteristics was critical in the success of the experiments. I will return to the implications of this decision shortly. He chose seven characteristics as the basis for his hybridizations: the shape of the seeds, the colour of the endosperm, the colour of the seed coat, the shape of the ripe pods, the colour of the unripe pods, the position of the flowers and the length of the stems.

Mendel crossed peas with different characteristics, for example wrinkled seeds with round seeds. He counted the numbers of plants with each characteristic in the subsequent (second) generation. He then crossed the second generation with each other, and counted the characteristics again. He then repeated the cross with the third generation and counted the characteristics in the offspring again. He continued the process until he was sure of the results. His experiments took eight years.

Mendel had no scientific training, and indeed the meticulous way in which he carried out and recorded the results of his experiments was unusual and unorthodox in scientists of the early nineteenth century.

From his results he could show that some 'characters', such as round peas, were dominant: if they were present on a parental plant, then they would appear in half of the next generation. Other traits, such as wrinkled peas, were recessive: that is they disappeared in the second generation, only to return in a quarter of the third generation. He also saw that the inheritance of one trait was independent from the inheritance of another.

From all of this he could deduce that the ovaries and pollen cells of the plants in some way contained the characteristics shown in the plant itself. In a hybrid, formed from plants with two different characteristics, the two forms would be represented in the egg and pollen cells in equal numbers.

Mendel had thus realized that there was a unit of inheritance for each characteristic of the plant. He also recognized that there were two copies of each unit of inheritance. He had no idea of the existence of DNA or chromosomes, and could not possibly have imagined the miniscule scale of the molecular events that determined the size and shape of his peas. Nevertheless, from his simple experiments Mendel had recognized the true nature of the gene.

Most science proceeds in small increments, so that several individuals reach similar conclusions in different places at about the same time. For Mendel's discovery there was no precedent; it was a *coup de foudre* from a clear sky.

Mendel presented the results of his experiments to the Natural Sciences Society of Brünn at two meetings in February and March of 1865. He then wrote a description of the work. This monograph gave simple mathematical rules for heredity, which are still found, almost unchanged, in today's textbooks of genetics. The monograph was published in the transactions of the Brünn Society, under the title *Versuche über Pflanzen-Hybriden* (Experiments in Plant Hybridization). The article was distributed to major libraries in England and the Continent. It was treated with resounding indifference, because no-one who read it had the remotest idea what it all meant. The use of mathematics was completely foreign to biologists of the time, and Mendel was perceived as an obscure provincial who was unlikely to have anything important to say.

Mendel wrote a series of polite letters to a leading biologist of his time, Carl von Nägeli of Munich. The letters are a detailed exposition of Mendel's theories. Von Nägeli was supercilious in his replies, but Mendel was grateful that the great man had even deigned to answer his letters. Von Nägeli suggested that Mendel turn his attention to another plant, called hawksweed. Due to a quirk in the life cycle of this weed, which was only recognized in this century, these experiments were failures. Mendel published no further scientific papers.

Mendel became prior of his monastery, and like many academics burdened with administrative responsibility, he gave up his experiments. A gentle, modest man, it seems likely that he himself attached no great importance to his results. He died in 1884 of kidney failure, after a 'long, severe, and painful illness'.

Mendel's paper was rediscovered in 1900, by which time other botanists were studying the transmission of various traits. Among those who read the paper was William Bateson, a formidable botanist, and an eminent figure in British biology at the turn of the century. He was the inventor of the term 'genetics', although the term 'gene' did not appear until 1909. From his own studies Bateson had seen plant characters to be inherited, and so Mendel's work struck an immediate chord. Bateson publicized Mendel's findings with missionary zeal.

There then followed an enormous controversy, the first to wrack the new science of genetics but certainly not the last. The problem was that many human and plant characteristics are not inherited in the same clear-cut way as the shape or colour of peas. We are not, for example, either tall or short, or fat or skinny. Mendel himself had ignored many characteristics in his peas that he could not classify with any precision. This decision was inspired: if you are to make sense of something complex, begin only with things about which you may be sure. However, this was perceived by his critics as examining special cases that had no bearing on the mainstream of heredity.

William Galton (see 'The New Eugenics', page 186 ff.) was born in the same year as Mendel, but was still scientifically active and influential at the end of the century. Galton had been measuring all that he could about human beings, and attempting to establish rules of heredity from these measurements. His work led to the 'biometric' school of

Law of Ancestral Heredity

heredity, the foremost proponent of which, after Galton, was a mathematician called Pearson. Pearson invented many statistics that are in common scientific use today. Galton, Pearson and their friend Weldon believed that heredity operated by a blending of such things as height and weight, or, notoriously, intelligence and ability. This theory was known as the 'Law of Ancestral Heredity', and its followers as the Ancestrians.

The details of the debate, and the skulduggery and upstaging that raged in the Royal Society committees, is not interesting now. We can, however, get some of the tone of the debate from Bateson's public statement: 'The imposing Correlation Table into which the biometrical Procrustes fits his arrays of unanalysed data is still no substitute for the common sieve of trained judgement.' And we can catch Weldon's ire when he describes '...the cumbrous and undemonstrable gametic mechanism on which Mendel's hypothesis rested'.

This behaviour is a fine example of scientists' failure to accept new ideas, or to modify their theories to fit new information. Pearson's main interest was statistics: he saw everywhere statistical distributions of inherited traits such as height and cranial capacity. He therefore had a huge investment in the biometric approach which was impossible to give up. Bateson, on the other hand, had written an ill-received book in 1894 called *Material for the Study of Variation*. This laid stress on the discrete nature of inherited characteristics and was savagely attacked by Darwinists. Mendel's experiments showed that Bateson was right, and thus Bateson was committed to Mendel's cause.

The argument and the resulting polarization into opposing camps set genetics back several years. It was not until 1918 that the Cambridge statistician Fisher showed that the biometric and Mendelian schools were entirely compatible with each other. It is now obvious that height is not inherited as a single character because it is due to interactions between a multiplicity of genes. The protagonists in the argument were not, however, interested in the synthesis of their disparate theories into a cogent whole.

The Mendelian view was given a huge lift when, in 1901, Garrod described the congenital disorder of alkaptonuria. This rare error of metabolism causes urine to turn black. In mid-life, sufferers develop progressive arthritis. Garrod said that alkaptonuria was most often seen in inbred families. In 1902 Bateson and Edith Saunders read his description. Because the disease affected the children of first cousins, Bateson and Saunders deduced that the cousins were likely to share the same abnormal gene. This meant that alkaptonuria was a recessive disorder. This was the first recognized example of a Mendelian disorder in man.

After Mendelism was accepted, it was only a short step to match Mendelism to the chromosomes. Boveri showed in 1902 that chromosomes did not all carry the same genetic material. In 1903, Sutton discovered that chromosomes were paired, and that one member of each pair came from the father and the second from the mother. The exceptions to this were the sex chromosomes. Females had a pair of X chromosomes, whereas males had only one. The Y chromosome, which partnered the male's X, was discovered later. In the sperm and in the ova, the number of chromosomes was exactly half normal. Sutton suggested that the chromosomes could '...constitute the physical basis of the Mendelian law of heredity'.

Thus it was established that there were two copies of each gene, one from each parent, and that the genes were carried on the chromosomes. Bateson, his status and audience swollen by his debates with the Ancestrians, virulently and completely mistakenly opposed this new chromosomal theory of heredity.

The chromosomal theory was carried forward by an American geneticist, Thomas Hunt Morgan. Around 1900 Morgan was opposed to Mendelism. Consequently Bateson disliked him, and wrote that 'Morgan is a thick head.' By 1910, however, Morgan had accepted Mendel's propositions, and was studying the genetics of the fruit fly. Morgan had realized that fruit flies are ideal for genetic study because they breed several times a year and have hundreds of offspring. Even today the same fruit fly species are still giving critical insights into genes and their function.

Morgan looked for abnormalities in the flies and studied their inheritance. He invented the term 'mutant'. Fruit fly mutations such as an extra set of wings, long bodies or two heads can be found simply by looking at the flies with a magnifying glass. In 1910 Morgan found a fruit fly that had white eyes instead of red. This meant that the gene for making red eye pigment in these flies was mutated, or broken. The remarkable thing about the white eye mutation was that it was found only in males. Morgan deduced that the white eye mutation must be on the X chromosome. It was only seen in males because they lacked a second X chromosome to compensate for the defective gene for eye colour. A very similar thing had been noticed in humans, where hereditary colour blindness was usually found in males.

Morgan deduced that the genes for sex and for eye colour must be inherited together. He surmised that this was because they were close to each other in the genetic material. Morgan went on to find other sex-linked mutations. Like most successful scientists, he attracted clever people to work with him, in particular three brilliant young scientists called Bridges, Sturtevant and Muller. In later years E.B. Wilson was to write archly to the British geneticist Darlington that 'Morgan's three greatest discoveries...were...Bridges, Sturtevant, and Muller.'

The other sex-linked mutations were inherited together as a group, leading to the idea that they were all carried on the same chromosome. However, pairs of mutations sometimes became separated from each other in their passage down the generations. This was attributed to exchange of genes between the two members of a chromosome pair. The chromosomes of germ cells (leading to ova and sperm) were seen to twist around each other just before the cells divided. This could be the moment that genetic material was exchanged between chromosomes. Due to the primitive microscopy of the time, they could do no more than guess at the chromosomes' behaviour. Bateson, who disliked microscopy because he was no good at it, seized upon this weakness in the explanation of Morgan's findings.

A.H. Sturtevant was still an undergraduate in Morgan's laboratory when he found that some pairs of mutations separated from each other more frequently than others. He realized that this meant that some genes were close together and others further apart, and that he could deduce the order of genes on the chromosome. He had invented the genetic map.

By 1916 Morgan and his team had found more than a hundred mutations, and had bred half a million flies. They had discovered that the mutations formed four sets – that is, there were four groups within which the mutations were associated with each other. The number four happily coincided with the number of fruit fly chromosomes.

Morgan and his three lieutenants wrote a book, *The Mechanism of Mendelian Heredity*, which a mischievous editor of the journal *Science* gave to Bateson to review. The review runs to seven pages. Its tone is ironic and mocking, but never totally dismissive. He attacks the chromosomal theory:

> But we know that the number of genetic factors in various types of life greatly exceeds the...number of chromosomes.... At this point we meet the first of the far-reaching suggestions which Morgan offers...

Later in the review Bateson attacks Sturtevant for his theory of linear arrangement:

> Without insisting too much on the point, we can not avoid noticing that this complex web of theory is so exceedingly elastic as to be capable of being fitted to a framework of cytological fact, the converse of that for which it was designed.

He then attacks the numbers given in the book:

> Meanwhile the data look so intractable that a doubt has sometimes arisen whether the account here given may not be a consequence of some radical misunderstanding of the author's meaning.

and questions the interpretation of other results:

> The machinery for dealing with uncomfortable cases is extraordinarily complete.

The explanations of the 'uncomfortable cases' are actually elegant and correct interpretations of difficult genetic phenomena. At the finish of the review he praises

> ...the great extension of genetic knowledge to which it has led – greater far than has been made in any one line of work since Mendel's own experiments.

Why is it important after so many years to read Bateson's comments? For two reasons. Firstly, Bateson's motivation, an unreasonable prejudice against a new and difficult theory, is alive and well today. Secondly, it shows how difficult the birth of new ideas can be. At the time of a new scientific discovery even the scientists involved are often not sure that they are right. Indeed, it may be years before they can be vindicated or corrected.

Morgan's name is now immortalized in a unit of genetic distance, the centiMorgan, while Bateson is largely forgotten. Since Morgan, linkage maps between fruit fly mutations have been painstakingly built up over many years.

This kind of work is extremely tedious. Each fly with a newly discovered mutation has to be bred with flies with each of the previously known mutations. The numbers of both mutations are counted in the offspring to test for linkage between them.

This simple recipe for madness was continued by mouse geneticists: mice with mutations were crossed with other mutant mice. In response perhaps to the monotony of this approach, the mouse genetic literature abounds with the exotic names of mutations. 'Satin' and 'frizzy' are relatively straightforward, presumably describing an aspect of the coat, and 'shaker' and 'short-ear' might be characters from Tolkein, but 'umbrous', 'mottled agouti' and 'varitint-waddler' almost defy imagination.

Because the mouse genome is so much bigger than that of the fruit fly, it might be necessary to cross a mutant mouse with a hundred other different types of mouse before a gene could be pinpointed to a particular point on a particular chromosome. If this sort of experiment is difficult in mice, in humans it is quite impossible.

The simplest cases in humans are those illnesses which, just like the fruit fly's eye-colour, are sex-linked and can thus be recognized as due to a mutation located on the X chromosome. Haemophilia and muscular dystrophy are the best-known examples of sex-linked disease in humans.

The X chromosome, however, contains only a few percent of the human genes, and only a few percent of human diseases. Other genetic illnesses have been traced by an abnormal appearance of the chromosome. Down's syndrome, for instance, is due to the abnormal presence of three chromosome 21s instead of two. In muscular dystrophy, pieces of the X chromosome are frequently seen to be missing. This type of information can be very helpful when it is present. Sadly, however, most genetic mistakes are not on the scale of a fractured chromosome, but are instead due to changes in one or two letters of the genetic code. To find the site of genes causing human disease some much more flexible mechanism had to be found.

THE MODERN MAP

Mechanism for mapping human genes was invented by David Botsein

According to the folklore of American genetics, the mechanism for mapping human genes was invented when David Botstein was seized with the afflatus at a ski resort in Utah. Botstein is another very clever American. A large man with black bushy eyebrows, he demands the attention of anyone in the same room, letting fly with an unremitting barrage of questions, ideas and answers. At the time of his ski trip, Botstein worked at the legendary Massachusetts Institute of Technology, a high temple of the brilliant best of American science.

What Botstein saw on the mountain top was that minute differences in the DNA from different individuals could be used as markers from which to make a map. He wrote a paper in 1980 with Ray White and Mark Skolonik, of the University of Utah, and Ronald Davis of Stanford. The paper outlined a strategy for building such a map. On my side of the Atlantic there is a feeling that Botstein was not the only person to see the possibilities. Walter Bodmer and Ellen Solomon were suggesting a similar map as early as 1979. Nevertheless, Botstein and his co-authors had laid out with remarkable prescience the ground rules and methodology for making a comprehensive genetic map.

Their paper set out modest aims. The map could be used for genetic counselling, and 150 markers would suffice to cover all the chromosomes. The idea of using the map to hunt disease genes was not yet explicit. The authors could not imagine the thousands of markers that the map would hold only fifteen years later. They did not even know if they could find the 150 markers needed for a basic map.

The markers that Botstein and the others wanted to use for their map were RFLPs (pronounced 'rifflips'), which stands for 'restriction fragment length polymorphism'.

To explain, polymorphism is the key to genetics and to gene hunt-

Markers for the Map

ing. The word simply means 'having different shapes'. Very many genetic traits are polymorphic: the colour of hair and skin, height, the shape of the nose, and so on. These polymorphisms of appearance are due to polymorphisms of the genes that control them. The number and the organization of your genes and the number and organization of the genes in the person sitting next to you on the train are identical, but the genes themselves are slightly different. His genes for the size of muscles are different to yours. They are not greatly different, and the proteins they make are very similar, but there is something in the terminal exon of one of his genes which determined that his muscles should be larger than the average, and another genetic polymorphism somewhere has given him that unfortunate facial tic.

To look at the huge variety of the human race is to realize how polymorphic many of our genes must be. The degree of polymorphism permitted by evolution depends on the function of the gene. Some genes that are absolutely critical for development, such as those that operate during the growth of an embryo, are not polymorphic at all, and do not vary even between species. With other genes nature has had freer play, as long as a variant gene has still produced something that works.

In the vast tracts of DNA that do not code for genes, the limitations on polymorphism are even fewer. It is an evolutionary disadvantage if a mutation completely ruins a gene, but a similar mutation in a non-coding area will usually not matter at all. It therefore follows that mutations and polymorphisms are much more common in the DNA that does not contain genes. The most common polymorphisms affect only a single base or a pair of nucleotides in a twin strand of DNA. It was these ubiquitous 'point polymorphisms' that Botstein and his co-authors wanted to use as markers.

They proposed that the polymorphisms be detected by 'restriction enzymes'. These are the chemicals that molecular biologists or geneticists use to cut DNA. DNA-cutting enzymes are made by bacteria, which secrete a staggering variety of toxic compounds, some of which can cause human disease, as part of their war effort; they must battle to the death with other bacteria for the space and food they require to multiply. Restriction enzymes are part of the bacterial armamentarium, dismembering foreign DNA.

Some restriction enzymes cut DNA only when and where they can find a certain sequence, for example T G C A. This specificity prevents the demolition of the bacteria's own DNA. Some enzymes recognize a four-base-pair sequence, others six or more. A 'four-cutter' will, on average, find its particular sequences and cut the DNA once in every 256 DNA bases. Whenever a polymorphism or mutation is in a restriction site, so that, for example, T G C A becomes T G A A, then the restriction enzyme will not cut. The presence of the RFLP, and thus the missing cut, results in different lengths of fragment after treatment with the enzyme. Thus the term 'restriction fragment length polymorphism'.

Cutting genomic DNA gives billions of fragments. A RFLP at a particular site on a particular chromosome therefore has to be singled out. The RFLP is found by a 'probe', a piece of DNA that uniquely matches the sequence on that part of that chromosome. A probe will always find the same place from the same chromosome. A probe is one example of a genetic marker, although, as we shall see, many other kinds of markers exist.

Once he had decided RFLPs could signpost the human genome, Ray White in Salt Lake City began to assemble a team of gene mappers. His success in this is measured by the present prominence of several of the people that he trained: Lathrop, Nakamura, and Lalouel have all made their mark in international genetics.

Human mapping follows the same principles as the genetic mapping of T.H. Morgan's fruit flies. Morgan and his team collected mutations. They observed how the mutations were inherited with respect to each other. The mutations called 'Y', 'W', 'V', 'M', 'R' and 'Br', for example, were consistently inherited together more often than would be expected by chance. Morgan's team found that mutations fell into four groups which shared some inheritance. These 'linkage groups' they correctly attributed to the physical location of the mutant genes on the four fruit fly chromosomes. The strength of the co-inheritance of two mutations could be used to infer the distance between their genes: weakly linked genes were likely to be further apart on the chromosome than strongly linked genes. Morgan's team was also able to deduce the order of the mutant genes on each chromosome. Thus, for example, as W was linked closely to Y and V, but Y and V were not closely linked to each other, W was likely to be between the two. White's team collected RFLPs instead of mutants, and began to piece the RFLPs together into a map of the 23 human chromosomes.

It is not possible to breed populations of human beings for mapping experiments. Gene mapping has to be carried out on the nearest natural equivalent: very large families. The ideal family for mapping has two parents, about eight children and all four grandparents. Such large families are difficult to find, except in Utah, the home of the

Mormons. From their workplace in the capital of Utah, Ray White's team found fifty such families.

DNA for most experiments in humans is taken from white blood cells. A syringeful of blood is all that is needed. As well as taking DNA from the families for its own use, White's team had passed white blood cells on to CEPH, the Centre d'Etude de Polymorphisme Humaine, whose headquarters are near L'Hôpital St Louis in Paris, away from the tourists. Here in the real Paris are cafés where you can drink alcohol at seven in the morning, rumbling metros, shops stuffed with food, a gracious canal, and trails of dog urine running every few yards down the pavement.

The purpose of the CEPH, led by Jean Dausset, a Nobel laureate, is to store and distribute DNA and mapping information from large families. This allows anyone who wishes to map a part of the genome to use the same families as everyone else. The advantage is that any marker, from whatever source, can be ordered with respect to all the others. Without these 'reference families' mapping would be very inefficient; new mappers would have to test previously mapped markers on their own families, to orientate themselves.

After Ray White's group had started on their map, they were joined by unwelcome competitors. Collaborative Research was a Massachusetts biotechnology company which wanted to cash in on the exciting possibility of finding the genes that cause disease. Their scientific effort was led by Helen Donis-Keller, an established geneticist with a justifiably high reputation. At one time she was married to David Botstein.

As Collaborative were making a big mapping effort, they had free access to the CEPH family material. However, during 1984 and 1985 Collaborative were deeply involved in a scandal over the cystic fibrosis gene, about which more later. This made it clear that Collaborative's commercial motivation was overriding normal scientific practice. Relationships between Collaborative and Ray White grew increasingly strained. In the meantime, both had made enormous progress in generating a map, Ray White publishing his results chromosome by chromosome, and Collaborative keeping commercial silence.

In October 1987 Collaborative Research broke their silence and published a paper in the highly prestigious journal *Cell*. The title of the paper was 'A genetic linkage map of the human genome', which is not unpretentious. There were 33 authors to the paper, 28 of whom were from Collaborative. The paper was announced with great publicity. It described 404 markers, of which 306 had been identified by Collaborative, who claimed their map covered 95% of the genome.

Ray White and his team greeted the paper, and the press conference that accompanied it, with anger. The 'Flap about the Map' was gleefully described in another leading journal, *Science*. Ray White claimed the Collaborative map was incomplete. Of Collaborative's own 306 markers, 60 were on chromosome 7, reflecting their abortive efforts to find and patent the cystic fibrosis gene on that chromosome. However, there were only two markers on chromosome 14, and only five each on chromosomes 19, 21 and 22. There were also many holes in the map, regions with no markers at all. Ray White pointed out that his group had 470 markers, and that they had mapped them on 60 families, compared to the 21 families used by Collaborative.

These arguments may seem typical of academic bickering, but rather more than scientific precedence was at stake. The real issue was Collaborative's motive in publishing, and the brouhaha that went with it. Collaborative were beasts in the genome jungle. They wanted money. Collaborative would only release their probes if scientists signed a binding commercial document, in contrast to the Utah group who distributed their markers (or, more accurately, the probes that matched them) freely. Collaborative's hope was that scientists looking for genes would use their markers, and that any success in finding a gene would be reflected in Collaborative's balance sheet. The president of Collaborative, Thomas Osterling, had spoken in the past of filing patent applications 'for the probe and for any other probe between that one and the gene'. More famously, the Chief Executive Officer, Orrie Friedman, had said, 'We own chromosome 7.' Even if Collaborative never found a commercially important gene, the publicity could be expected to give its share price a substantial boost.

The conflict between commercial interest and academic science has been a recurring theme in genetic research. In 1987 Walter Gilbert of Genome Corporation laid out ambitious plans to patent any part of the genome that his company sequenced. In justification, people were talking about 'the bucks to ethics ratio'. In 1991 Harvard and the National Institutes of Health sparked off a big row by patenting c-DNA (gene fragments) that they were sequencing at random. They claimed rights to any gene whose sequence they had described. This was despite the fact that the sequence might be incomplete, including only part of the gene, and that they might have no idea at all of its function. The Medical Research Council in the UK reluctantly followed suit, to protect their own findings.

The c-DNA patent was mercifully not granted. To date, none of these ambitious and massively greedy patenting schemes has been successful, but if such a scheme ever thrives, catastrophe will result. This is not because making money from research is intrinsically sinful, it is not; rather, unbridled commercial interest blocks the free flow of information. The highest levels of science feed on the rapid communication of information, and in genetics, the transfer of materials such as probes. Results may not be published until a year after the initial discovery has been made. This delay is as much due to the need to check and extend the results as to publishers and their reviewers dragging their feet. Patenting does not take long, but commercial interests want to protect their investments by gaining 'lead time' over their rivals. They can then be well ahead on the next step before rivals are aware of their results.

If every research group feels obliged to keep all their results secret for as long as possible, then not only will their competitors suffer, but also their potential collaborators. If I am mapping a particular gene on chromosome 1, and someone in Germany is mapping another separate gene that is close by, then it makes sense to exchange results. It also makes sense to exchange any markers that we might have. If it takes an experienced worker a year to produce a set of map points, then he can save everyone else that year's effort by allowing free access to his markers. The pay-off for this altruism may be at best co-authorship on the collaborator's paper, but would usually only rate an acknowledgement. Greed and patents, however, mean there will be no collaboration, and everyone loses.

Happily, the technology in genetics is moving so fast that squabbling groups are often left behind. Almost as the fuss over Collaborative's map subsided, RFLPs were rapidly being superseded by better kinds of markers.

RFLPs have a severe disadvantage. Although common, they only come in two states: they are either present or absent. This can be thought of as two possible numbers, 1 or 2. Even if half the people in a population are 1, and the other half are 2, many marriages take place between people who are both 1 or both 2. It then follows that their childrens' RFLPs will also be all 1 or all 2. In many families, therefore, there will be no pattern with which to compare the inheritance of other map points.

The number of families and markers that need to be typed are reduced when the polymorphisms not only correspond to 1 and 2, but also to 3, 4, 5, 6 and so on. A new class of 'hypervariable' (HVR) markers were discovered, with much more polymorphism than simple RFLPs.

In an HVR a sequence of DNA, often about thirty nucleotides long, repeats itself hundreds of times. They are known as VNTRs, which stands for 'variable numbers of tandem repeats'. 'Tandem' is jargon for one repeat following another on the DNA.

It is possible that crossing over and recombination between chromosomes takes place within these repeating sequences. An analogy is with two strings of plastic pearls. Both break near the middle, and the opposite pieces exchange with each other. The core repeated sequence of the VNTR is like one pearl. Sometimes the number of pearls which are changed are not the same, and after many exchanges the strings can become quite different in length. VNTRs too have become different in length, and instead of one or two variants, there can be ten or more. For a gene mapper this is ideal, as marriages will nearly always be between people with different numbers of repeats.

A few of these highly variable regions had been found in proximity to well-studied genes. One such gene was for a part of haemoglobin, the blood pigment that carries oxygen. A second was the gene for myoglobin, another oxygen-storing pigment found in muscles. Alec Jeffreys, at the University of Leicester, was studying the repeated sequence from the myoglobin VNTR. Out of curiosity, he tested to see if there was anything similar elsewhere in the genome. He found a wonderful and exciting thing.

After some elegant cookery, Jeffreys had made a probe that identified the repeated central sequence of the myoglobin VNTR. To his surprise he found a whole family of similar VNTRs, and then found other VNTR families. Each family had many members, scattered through the chromosomes. Within families, the repeated core sequence was shared, although it differed between families. Jeffrey's probe, and others identifying different repeated sequences, could therefore show many VNTRs on a single sample of DNA. With his probes he could make a pattern of DNA bands rather like a supermarket bar code. Each band came from a different place on a different chromosome. The really extraordinary thing about this bar code was that no two individuals shared the same pattern of bands.

Jeffreys immediately realized that he had a means of identifying individuals by their DNA. This identification was thousands of times more accurate than by any print left by a guilty finger. For this technique he coined the term 'genetic fingerprinting'. Jeffreys could also tell from the DNA whether two people were related or not, because

Identification through DNA

each parent contributes half of an individual's genetic fingerprint, just as each parent contributes half of their child's genes.

This discovery has altered forensic science forever. If a criminal leaves a sample of DNA at the crime, and a sample may be as insignificant as a single hair, then that criminal can be matched to the hair by genetic fingerprinting. The identification of rapists and murderers is now sure when before it was often impossible.

Between 1986 and 1992, 100,000 of these tests were carried out in the UK. These included paternity testing on 20,000-year-old mummies, the identification of the remains of Joseph Mengele, who died in 1976, and the identification of the family of the Tsar, who died in 1917.

Genetic fingerprinting has greatly affected the outcome of immigration cases. Before fingerprinting, it was commonplace for the families of Asian UK residents to be denied entry because they could not prove that they were blood relatives of the UK residents. The assumption by the authorities was always that the applicants were lying. Now, just a few years after the introduction of genetic fingerprinting, the numbers of immigrants are falling off steeply. This is not because the fingerprint has detected all the liars, but rather because it has shown that most of them were telling the truth. The numbers are decreasing because the end of the queue has at last been reached.

A geneticist's rule is *Pater semper incertum est*: only maternity is certain. As genetic typing by RFLPs and fingerprinting has been more widely applied, it is evident that a consistent proportion of most populations do not belong to their supposed fathers. As many as 5% of us are the result of unadmitted liaisons. Perhaps it is all down to the DNA, seeking new combinations of genes by whatever means it can. It is a moral relief to realize that other apparently monogamous species have behaved the same. Konrad Lorenz used to wax lyrical about the faithfulness of wild geese, who choose a partner for life. Sadly, genetic fingerprinting shows them to be as capable of landing overnight in the wrong nest as humans.

DNA fingerprinting has come under attack in the courts, most notably in the United States. Part of the attack has been against the statistical interpretation of results. A jury will understandably find it difficult to apply statistics to the question of guilt or innocence; are odds of a thousand to one, for example, sufficient to send someone to prison for a life sentence? Genetic fingerprinting often gives odds much more certain than a thousand to one, but any uncertainty can be exploited by a good defence lawyer with a sympathetic expert witness. On the other hand, genetic fingerprinting is deservedly criticized when the test has not been performed properly: the bar code ~~may be~~ smudged, or bands may be thought to be the same when they are only similar. Here science can blind judge and jury into accepting innaccurate results.

Jeffreys has dealt with both difficulties by refining the fingerprint further. His most elegant method of fingerprinting gives a result in a binary code. This is like a huge number that is unique for everyone, although half the number still derives from each parent.

Jeffrey's original probes identified bands from all over the genome. By finding the sequences that were on either side of the repeated core motif, he later derived markers from these bands that were specific for particular chromosomes. These probes, though wonderfully helpful to gene mappers, were limited in number. Other repetitive sequences were found, and other types of hypervariable region (HVR). A brilliant Japanese molecular biologist, Yusuke Nakamura, developed a systematic method for finding new HVRs. He spent five years with Ray White in Salt Lake City, in the process becoming a gene-mapping legend. Rumoured to have slept only fifteen hours a week in his first year, he had a zest for work which was matched with a ferocious organizational ability, and hundreds of the best markers in the genome were the result of his efforts.

The ultimate marker came with the discovery of microsatellite repeats. These also contain many repeats of the same genetic sequence, except that the sequence is only two or three base pairs long. The most common of these is C A, repeated between ten and thirty times. Microsatellites are extremely common; there may be 60,000 of them in the genome. In 1993 Mark Lathrop, the former pupil of Ray White, together with Jean Weissenbach, had made a 'second generation' map of 800 microsatellites. By the end of the year the number had grown to several thousand. There is no point now in making more markers, other than to fill in gaps: the genetic map can be said to be nearly complete.

However, completing the map does not mean that all the genome has been discovered. Even in 1994 only 4,000 of the 100,000 human genes have been tied down to particular chromosomes. The number of disease genes localized is much smaller: only about 900 disorders have been mapped to a particular chromosome, and only 150 genes have been sequenced and found to contain mutations. This is tiny compared to the 3,500 genetic disorders known to exist.

The completion of the map means that anywhere in the genome can be reached from a known map point, in a continuous coverage of all

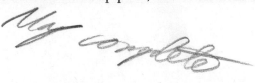

chromosomes. In practical terms, any gene or genetic illness can be positioned in its correct place on its chromosome. The stage is now set for large scale gene hunting.

Before embarking on the journey into the genome, we should understand how a molecular geneticist can capture a gene in a test tube. A gene is captured when it is cloned, a process which, in fiction, has been used to reproduce a host of sinister creatures, from dinosaurs to Nazi leaders. In fact, though cloning is applied to less virulent forms of life, it is still an amazing process.

Cloning was invented only twenty years ago, in San Francisco; just one of a range of scientific achievements which will ensure that, when the history of our age comes to be written, California in the second half of this century will be recognized as a time and place equivalent to Britain during the industrial revolution, when developments in engineering and science brought about changes which altered the lot of humanity for ever.

Two separate lines of research led to cloning. The first came from the study of bacteria. Penicillin was the first antibiotic used to treat an infection – in the Radcliffe Infirmary, Oxford in 1941. After the war, the use of antibiotics became almost universal and, for the moment, humanity is winning the battle against infection. Because bacteria differ in their sensitivity to antibiotics, physicians have developed tests for bacterial resistance. Resistance to an antibiotic is a genetic trait, carried in the bacterial genes. Microbiologists routinely grow bacteria from infected patients on plates of a solid nutrient. The bacteria are spread, so that they grow all over the plate. Spots of antibiotic are placed on the plate and sensitivity is revealed by a clear zone, with an absence of bacterial growth around the spot. The microbiologist can then tell the doctor how best to treat his patient.

Bacteria only have one chromosome, so it was assumed that all their genes resided in that single aggregation of DNA. The study of infections, however, showed that bacteria could transmit their resistance to antibiotics to other bacteria. This was not through the usual mechanisms for swapping genes, known more simply as sex. How the genes were transferred between the bacteria was a mystery.

The puzzle was solved by the discovery of 'plasmids'. These circular pieces of DNA were originally known as 'R (or resistance) factors'. These DNA loops contain only two or three genes. They are completely separate from the genes on the bacterial chromosome. Though a bacteria has only a single chromosome, it may contain many plasmids. Plasmids have the property of entering bacteria quite easily if the bacteria are damaged or stressed. They therefore afford bacteria an easy way of spreading the genes for antibiotic resistance.

Stanley Cohen was working at Stanford University Medical School, at the beginning of the 1970s, when he found a mutant R factor that had lost its resistance to the common antibiotic tetracycline. Cohen showed, by electron microscopy, that the tetracycline resistance gene in the plasmid had been interrupted by a piece of DNA running in the reverse direction. This led him to undertake other experiments in which he tried to insert other pieces of DNA into the plasmids or to remove them. Unfortunately, the only means of fragmenting DNA was literally to break it, shearing it into pieces with the equivalent of a food processor. From the subsequent tangle of fragments, he was sometimes able to reconstitute plasmids with interesting functions, but the process was extremely inefficient.

In the nearby San Francisco Medical Centre, Herb Boyer was studying restriction enzymes – the same enzymes that would later make RFLPs for the genetic map. Boyer was analysing an enzyme from the most common bacteria in our intestines, *Escherichia coli*, known more affectionately as *E. coli*. *E. coli* makes many restriction enzymes, but Boyer was interested in the first to be found, called Eco R I. In 1972 Boyer and his team used a very primitive form of DNA sequencing to analyse where Eco R I cut the DNA. They proved that Eco R I always cut the same sequence, which ran T G A A T T C T in the forward direction.

Since DNA comes in two matching strands, where the base T always matches A and C always matches G, the sequence where Eco R I cut the second strand of DNA read A C T T A A G A, and the two strands together looked like this:

..T G A A T T C T..

..A C T T A A G A..

However, 'forward' on the second strand of DNA is in the opposite direction, that is it reads from right to left instead of from left to right. Reading the second strand from right to left gave the sequence A G A A T T C A, the middle six letters of which were identical to the top strand: in Boyer's words, when he and his co-workers published their results, 'The most striking feature of the sequence is its symmetry'.

If the symmetry was interesting, what was fascinating was where the enzyme cut, which was between the G and the A A T T, like this:

..T G * A A T T C T..
..A C T T A A * G A..

The two cut ends therefore looked like this:

..T G   A A T T C T..
..A C T T A A   G A..

This pattern of cutting left the two fragments, each with a matching sequence of As and Ts open on one of its strands. These 'sticky ends' could bind back to each other, or to any other piece of DNA that had been cut with the same enzyme. Since 1973, many other restriction enzymes have been discovered in addition to Eco R I. Many cut in a similar way, recognizing palindromic sequences on the two strands of DNA.

A year later Boyer and Cohen together published the paper that ignited the genetic revolution. They opened plasmid circles with Eco R I. The enzyme only cut the plasmid once, so that it became a line instead of a loop.

Left to themselves, the two sticky ends would eventually close, reforming the loop. However, Boyer and Cohen also cut from other organisms DNA with Eco R I, which produced many fragments of different sizes, each with two sticky ends. On average, the fragments were the right size to contain whole genes. When the fragments and the open plasmids were mixed, a fragment was sometimes introduced into a plasmid circle. This artificial plasmid, containing a gene of 'recombinant' DNA, was known as a chimera or hybrid.

In Greek myth the Chimera was a monster with the head of a lion, the body of a goat, and the tail of a serpent, while the original hybrid was the offspring of a domestic sow and a wild boar. I prefer to think of the genome jungle as populated by the exotic and terrifying chimera, rather than by vigorous pigs.

Cohen knew, from his earlier work, that plasmids easily entered bacteria if they were leaky. It was simple to make them leak by heating them or treating them with salt. Boyer and Cohen showed that the chimeric plasmid, which they had made in a test tube, could be reintroduced into living bacteria. Once inside the bacteria the plasmid replicated like a normal plasmid, making multiple copies.

By good fortune, Eco R I cut into the tetracycline resistance gene on

Cohen's plasmid. Bacteria containing chimeric plasmids were no longer resistant to tetracycline. This immediately made it possible to detect which bacteria contained recombinant molecules, and which contained only the original plasmid.

In ideal conditions, warmed in an incubator and shaken in nutrient broth, bacteria replicate every few minutes. A chimeric plasmid, with five copies in a bacteria, will have produced ten copies after five minutes, and twenty after ten minutes, forty after fifteen, 320 after thirty, 20,000 after an hour and so on. By the next morning the bacteria in the broth contain billions of identical chimeric plasmids.

It is a simple process to take the bacterial soup and separate out the plasmids. The chimera can be cut by Eco R I to free the gene from the plasmid. For the first time a gene could be caught in a test tube, grown to any desired amount, and harvested.

Boyer and Cohen had hijacked the machinery of life. They had invented cloning. The first genes to be cloned into a plasmid were ribosomal genes, which are concerned with the manufacture of proteins within the cell. These were chosen for cloning because they are so common in cells, and easy to isolate. Other genes soon followed.

Genetic engineering is about the manipulation of cloned genes. Cohen's plasmid was very simple. It contained the gene for tetracycline resistance, and a sequence that allowed it to copy itself like a normal plasmid, and not much else. In the years since 1973 man-made plasmids have become more sophisticated. Cloning sites for many other restriction enzymes have been introduced, as have other genes that show whether cloning has been successful. Other types of DNA can also be used for cloning. They are known collectively as vectors. It is desirable that vectors carry as much DNA as possible. Plasmids only contain a few thousand base pairs. Another vector is a virus, called phage, which can carry ten times as much DNA as a plasmid. Recently, more synthetic vectors have been invented, the queen of which is the YAC. YACs are 'yeast artificial chromosomes', which are used to clone up to a million base pairs of DNA.

Very importantly, vectors were built with the machinery to express the cloned gene. Cloning a gene into an 'expression vector' means the gene can produce its normal product. The bacteria containing the vector then act as a factory for the protein coded by the imported gene.

The first medically important gene to be cloned and expressed was the human insulin gene. Before it was cloned, diabetics relied on insulin that had been painstakingly extracted from the pancreases of pigs and cows. Porcine and bovine insulin is not quite the same as human, which meant that some diabetics became immune or allergic to it, preventing it from working adequately. Cloning the gene meant that an infinite supply of pure human insulin was available.

Even today, the number of human cloned genes is limited and the number used directly in the treatment of disease is smaller still; but these few are spectacular successes. Erythropoetin, or EPO, is a hormone that induces the growth of red blood cells. Normally EPO comes from the kidney. People with kidney failure suffer from terrible anaemia, so that they feel desperately tired and weak. Dialysis, although effective in removing toxins from the blood, only partially improves this anaemia. Because it cures their anaemia, EPO has changed the lives of people in kidney failure. Its worldwide revenues are nearly a billion dollars a year. Growth hormone, used to treat short children, has incomes of a quarter that amount. TPA is a normal protein that encourages clots in the blood to dissolve. Recombinant TPA helps reduce the damage to heart after a coronary occlusion. It too generates an income of hundreds of millions of dollars per year.

When Herb Boyer realized what he had invented, he celebrated the discovery by going to a San Francisco bar. There he met a young lawyer, Robert Swanson. They fell into conversation, and Boyer told Bob Swanson that he had just invented cloning. Swanson liked the idea, and said that there should be money in it. Together Swanson and Boyer formed the first biotechnology company, now the mighty Genentech.

Genentech today has an income in excess of half a billion dollars a year. It spends half of this on research and development and its researchers publish as many papers in leading journals as those at many universities. The company headquarters sprawls over several acres, looking out over a serene San Francisco Bay. The car park contains many Porsches. At the end of every week there is a happy hour, where the Genentech team drink to the memory of the company's birth.

On the Genentech site is the manufacturing plant, where recombinant proteins are grown from genes. Human genes do not grow very well in bacteria, and so they are cloned into yeasts or mammalian cells. A brewer would therefore feel perfectly at home inside the Genentech production facility. There is the smell everywhere of yeast, a distinct improvement on the odours associated with E. coli. Vast stainless steel vats stretch up through the building. Inside each vat thousands of litres of nutrient broth are stirred and warmed and cosseted with gases to grow the cells and their recombinant proteins. The area is kept ultra-sterile: bacterial contamination would cause millions of dollars to be washed away down the drains. At the end of all this, the pure protein product from a ten thousand-litre fermentation may be measured in tens or hundreds of grams rather than kilograms. It is no wonder that recombinant proteins are expensive.

Genentech is the dream, the ultimate genetic experience. Many a gene hunter, setting out on his adventures, hopes at journey's end to find his own Genentech basking in the California sunshine.

CHAPTER 5

BIG GAME

IMPORTANT

The first aim of the genetic map was the better diagnosis of genetic conditions. However, as soon as it was obvious that a human genetic map was feasible, a much more ambitious idea was formulated: finding a mutant gene from the knowledge of its chromosomal location. This was the opposite of the 'traditional' way of associating a gene with a disease. Usually a gene had been cloned, then mutations had been found which could be related to disease. The new approach was to be called, slightly confusingly, 'reverse genetics'. Reverse genetics was 'reversed' because it began with the disease in families, moved to a position on a chromosome, and then to the gene itself. Reverse genetics is now called 'positional cloning', better describing the localization and capture of disease genes.

The major attraction of positional cloning was that it was entirely logical. Localization of the disease would lead inevitably to its gene, even if the nature of the gene was previously quite unknown. With a good genetic map and enough families, the area that must contain a gene could be reduced to about a million letters of the genetic code. Even though a gene usually consists of a few thousand letters of the genetic code, at the time, the early 1980s, it did not seem to matter that no-one knew how to dissect these from the other million bases.

Three illnesses were immediate targets for reverse genetics: muscular dystrophy, cystic fibrosis and Huntington's disease. A fourth condition, adult polycystic kidney disease (APKD) was soon added to the list.

These diseases all ran in family patterns that Mendel would have recognized in peas, or Morgan in his fruit flies. They affected substantial numbers of people. Mostly it was easy to tell who had the disease, although sometimes it was difficult to tell who did not. The diseases were known collectively as single gene disorders because it was assumed that they were all due to straightforward mutations in specific genes. No-one could have guessed then that the reality would contain so many humbling surprises.

Reverse genetics really began with Duchenne muscular dystrophy (DMD). DMD affects little boys soon after they start to toddle. The first sign is often an inability to rise from a crouch, so that children fall over in the playground. As the weakness persists, the muscles first grow larger to compensate, a process called hypertrophy. Later they waste progressively. Children with DMD are intellectually and otherwise completely normal. Heartbreakingly, by their mid-teens they are confined to a wheelchair. A few years later they are dead from a respiratory and cardiac failure.

Duchenne muscular dystrophy has a distinctive pattern of inheritance: boys are affected, and their sisters are carriers, exhibiting little or no sign of illness. This pattern means that Duchenne muscular dystrophy, like haemophilia, is sex-linked. A sex-linked pattern of inheritance means the gene causing the disease must be on the X chromosome, and so sex-linked diseases are also called X-linked. Boys have only one X to go with their Y. If that X is defective they are susceptible to disease. Girls remain well because they have a second X to compensate for the defect.

In 1980 there were two main kinds of X-linked muscular dystrophy, the Duchenne and Becker types. Becker dystrophy was milder, and sufferers lived relatively normal lives for many decades. The clinical differences between these two syndromes were taken to mean that they were caused by two separate genes.

Because muscular dystrophy was X-linked, the region of search was already reduced to the X chromosome itself. The X chromosome is 4% or less of the genome, a considerable assistance. Finding the DMD gene was greatly helped by a second phenomenom. Breaks were sometimes seen on the X chromosomes of boys with the illness. It was a very reasonable assumption that these breaks were the actual cause of the dystrophy; either they ran through the gene, or were close enough to interfere with its function.

Cytogenetics, the science of studying chromosome abnormalities, had been well established in humans for twenty years, and in plants for fifty. By piecing together the pattern of the breaks, a region could be defined that must contain muscular dystrophy. The first probable localization of the dystrophy gene by the study of abnormal chromosomes began in 1979, where the gene was tentatively placed on the short arm of the chromosome.

However, most cases of muscular dystrophy did not have visible breaks in their chromosomes. Also it could not be assumed that a particular break was causing the illness: the break might have been a chance association. To be sure of localization, many fractured chromosomes had to be studied. As Duchenne dystrophy affects only one boy in 5,000, even the big clinics were unable to find enough abnormal chromosomes to localize DMD accurately. Researchers were therefore obliged to collaborate, putting their natural competitiveness aside for the common good.

Studying chromosome breaks was part of the old genetics. In the mapping of Duchenne dystrophy the new DNA markers and RFLPs worked for the first time. This technology was pioneered in England in 1982, particularly by Kay Davies, who was then working at St Mary's Hospital with Bob Williamson. Williamson was later to apply these new techniques to the hunt for the cystic fibrosis gene. Kay has long blond hair, and on the podium at her first big conference struck one venerable professor by her resemblance to Alice in Wonderland. Appearances can mislead, because beneath the fair hair Kay has a mind like a steel trap. She is now indisputable queen of the X chromosome, and the advance of molecular genetics in the UK can be attributed in no small part to her abilities.

The evidence from cytogenetic and RFLP mapping steadily mounted, all of it apparently pointing to a region on the X chromosome known as Xp21. At the beginning of 1986, it seemed that finding the gene was only a few weeks away. Sue Kenrick, a post-doc in Kay Davies' Oxford lab, was able to remark in June, without the slightest suggestion of levity, of events 'way back in January'. Progress was fuelled by collaboration. The eventual cooperation between scientists searching for the gene was without precedent. One important paper in *Nature* in July 1976 boasted 75 authors. A cynic might argue that this number of people could only have worked together because they had to. No group alone could command sufficient funding or families to find the gene. Non-collaboration meant no progress. Whatever their motives, the science benefited enormously from their cooperation.

However, the arrival of the 75 author paper brought the search to an apparent standstill. The evidence from the fractured chromosomes showed that breaks a huge distance apart could cause the same symptoms. This distance was between two and five million base pairs: most known genes were a few thousand base pairs long. A series of highly unlikely events, involving breaks within breaks, had to be invoked to explain the findings. No-one really believed them. There was no way forward.

The impasse ended suddenly, with a result from Boston. Lou Kunkel was a leader of the cooperative effort, but had also pursued his own initiatives and his own ideas. This was part of the tacit agreement wisely reached between collaborators, that each laboratory should be allowed to follow its own leads, as long as this was not at the expense of other workers.

Kunkel's group had decided to try a different approach to finding the gene. Tony Monaco, who was working in Kunkel's laboratory, had made a 'library' from muscle m-RNA. m-RNA in a particular cell represents only the working genes, and does not include all the odd sequences that are found in DNA. For various reasons RNA is converted to complimentary DNA ('c-DNA') before a library is made. A library consists of millions of c-DNA molecules cloned into separate bacteria. A single bacterium, with its single clone, can be isolated and grown to produce billions of copies of one gene.

Monaco's library therefore contained most muscle genes. It also contained many thousands of the 'housekeeping' genes that are present in most cells of whatever type, to carry out their basic metabolism. The problem was to isolate the muscular dystrophy gene from among all the others in the library. Monaco knew that the muscular dystrophy gene was in the chromosome region Xp21, and he also knew that the gene was likely to be in his library. On its own, neither bit of information was sufficient to isolate the gene.

The answer was simple in concept, but extremely difficult in practice because of the arcane molecular 'cookery' required. Monaco took segments of DNA that had been isolated from the muscular dystrophy region of Xp21, and looked for a match with the c-DNA clones from the muscle library. The approach was successful. He found a c-DNA clone that matched the DNA from Xp21. He had found the muscular dystrophy gene at last. Reverse genetics worked. Now anything was possible.

However, the clone was only part of the gene and the whole structure had then to be painstakingly pieced together from different libraries and different clones. The gene that was revealed was bizarre: it was far bigger than any other known gene and a hundred times bigger than most common genes. It was so big that it even had other genes running inside it. The mystery of the map was explained: breaks that were two million bases apart could cause muscular dystrophy because they were both inside the same gene.

The DMD gene makes a protein called 'dystrophin'. Kunkel's group found dystrophin was missing from muscles of boys with the disease, although it was present in normal people. Dystrophin acts as an anchor between the contractile mechanisms in the muscle cell and the cell walls. Without dystrophin the contractile proteins tear away from the walls, causing progressive damage and eventually death of the muscle cell. Intriguingly, the gene was found to be expressed in tissues other than muscle, especially the brain, which suggests that other parts of the gene are expressed in these tissues and that the same gene has been used by nature to make more than one protein. The functions of these other proteins remain mysterious.

Becker was the first casualty of a gene hunt. He had described his syndrome in 1957, long after Duchenne's 1868 paper. So when it was found that fatal Duchenne dystrophy and mild Becker dystrophy were simply different mutations in the same gene, it was Duchenne who maintained his eponymous hold on history, leaving Becker without a disease to his name.

Before the DMD gene was found, no-one had dreamed of the presence or the role of a protein like dystrophin. The search for the muscular dystrophy gene was the first great gene hunt and showed that the whole improbable process of reverse genetics could be made to work, at least within the relatively safe confines of the X chromosome.

The future was rosy. Science-watchers waited for the next quick triumph of the new genetics: tracking down the gene for cystic fibrosis.

Cystic fibrosis (CF) is a disease of children. Children with CF are born with an inability to make normal secretions. This affects their sweat, which is very salty. The first sign of cystic fibrosis can be a mother's complaint that her baby's skin 'tastes different'. The abnormal secretions also affect the pancreas, the organ that makes enzymes to digest our food, preventing children with the illness from absorbing their food properly, and thus inhibiting their growth. Indeed it was originally thought that cystic fibrosis was primarily a pancreatic disease. Sometimes the child's intestines are blocked at birth. Males with the disease are usually infertile. Most of these defects are treatable; one, however, is not. The mucus in the lungs is intractably sticky.

Normal lungs are made of a many-branching tree of airways. As the branches narrow to invisibility, they end in billions of air sacs with a combined area the size of a tennis court. This huge area is necessary for oxygen to enter the blood stream, and for carbon dioxide to be blown off.

The air we breathe is not a simple mixture of pure gases. Any child who has looked at a beam of light in a darkened room will tell you that the air is full of dust, made up of many things. Flakes of human and animal skin, bacteria, pollen grains and fragments of mites are the most common pollutants in our houses. Outside, even in the purest air of the open country, each breath brings millions of particles into our lungs.

To cope with this detritus, the lungs and the nose are equipped with many defence mechanisms. The most important defence is the 'muco-ciliary ladder'. This has two parts. The first of these are cilia, microscopic hairs that line the airways of the lung. Cilia are covered with mucus, a complex mixture of proteins and water. Mucus also has a special immunoglobulin, or antibody, which binds to bacteria and other objects, keeping them in the mucus and protecting the lung underneath. The cilia beat, like a millipede's legs, carrying the mucus up and out of the airways.

The moving layer of mucus acts like a conveyor belt, bringing the inhaled debris away from where it can do harm. Eventually the mucus from the lungs is swallowed without us being at all aware of what is happening. Only when we get a cold or other infection, and the volume of mucus increases, do we appreciate its existence.

In children with cystic fibrosis the mucus in the lungs is thickened and viscous, and too heavy for the cilia to move. Instead of a flowing surface carrying bacteria away, the mucus forms stagnant pools. Wherever there is stagnation, bacteria can multiply, just as a backwater becomes choked with duckweed but a flowing stream remains clear and clean. When the bacteria multiply the other body defences, such as white blood cells, attack them; both the blood cells and the bacteria release highly reactive chemicals as they wage war with each other, infusing the mucus with toxic substances.

These toxic materials damage the airway lining, and when the damage exceeds the capacity for repair the result is irreversible scarring and fibrosis. In a matter of years the fibrosis has extended through the lung. The airways form distended irregular dilations, the cysts that give the illness its name. Infection in these cysts spills over into relatively normal areas of the lung, in turn predisposing them to more frequent and more severe infections.

One child in 2,000 is born with cystic fibrosis. There are 8,000 such children in Britain, 30,000 in the United States, and 70,000 in the European Community. At birth their lungs often appear normal. The chest X-ray shows the shadows of the ribs and backbone, the central heart and thymus, and the delicate ordered tracery of blood vessels and airways fanning out into transparent and pristine lung fields.

But, no matter how good treatment is, or how diligently a child submits to the twice-daily grind of intense physiotherapy, the repeated infections will gradually take their toll. The chest X-ray will change. The lungs fill with white scars and shadows, the heart and the blood vessels enlarge under the strain of pumping blood through the shrunken lungs, and the bones deform to match the lung they cannot protect. Finally heart and lungs fail completely.

Only twenty years ago it was rare for children with cystic fibrosis to survive beyond their early teens. Now it is commonplace for them to live well beyond twenty or even to thirty. This improvement in life expectancy is partly due to improved treatment with physiotherapy and antibiotics. Another important factor has been the recognition of the pancreatic failure that accompanies the lung illness; because their pancreas does not secrete sufficient digestive enzymes, children with CF, if left untreated, will suffer chronic malnutrition.

Cystic fibrosis runs in families, affecting brothers and sisters, and skipping generations. This is the pattern of a recessive genetic defect: one in which a child has to have inherited an abnormal copy of the gene from both parents, who were carriers. One quarter of the brothers and sisters of children with CF will also inherit the disease. One half of them will be carriers like their parents.

The number of carriers in the population is high: as many as one in twenty of us carries an abnormal CF gene. Why the gene should be so common is unknown. Cystic fibrosis is rare in people of African or Chinese descent, and is largely confined to peoples of European origin. But genetic mutations which are present in 1% or more of the population are usually there for a reason. It is possible that being a carrier of CF gives an advantage. In Europeans the most recent severe selective pressure on our evolution, or on our repertoire of genes, has been tuberculosis, which at the start of the nineteenth century used to kill one person in three. Before TB there was the bubonic plague. To be a carrier for CF may protect against one of these infections.

Other recessive genetic illnesses are common in certain populations because they defend against infection. The best known of these is sickle-cell anaemia. This causes severe anaemia and disease of the spleen and bones, and is due to an abnormality of the genes for haemoglobin, the red pigment that carries oxygen in the blood. The abnormal haemoglobin is called HbS. The unfortunate people who carry two HbS genes are very unwell. Those with only one abnormal gene have some HbS in their red cells and are mildly affected.

Genes that cause severe diseases like sickle-cell anaemia usually disappear from a population because people with them do not have as many children as people without. However, the malaria parasite finds it hard to survive in red cells with HbS and this protection becomes very important where malaria is common. In parts of Africa, one person in three carries the abnormal haemoglobin molecule. Although people with severe sickle-cell disease are common in these areas, the lives saved from malaria outweigh the lives lost to anaemia.

In the Mediterranean region there is another common disease of haemoglobin, called thalassaemia. Like sickle-cell anaemia, it can cause lethal illness but also protects against malaria and, again, it has been maintained in the population.

Recessive diseases of fat storage cause severe illnesses that are fatal after a few years of life. They have their maximum incidence in the Askenazim, who derive from eastern Europe. The prevalence of the disease in their descendants matches the rate of tuberculosis in the cities of their origin. It follows that fat storage mutations were probably protective against TB.

It may not matter too much if evolutionary pressure or an accident of fate led to the high frequency of CF carriers in Europeans . What is most important is that the effect of two abnormal CF genes is devastating. Before reverse genetics, the abnormalities noticed in sweat suggested some possibilities but, despite very substantial research efforts, the cause of CF remained a mystery.

This is why Botstein and Bodmer's articles (see 'The Modern Map', page 31 ff.) attracted so much attention among those concerned with the disease. If it was possible to find the CF gene by positional cloning, then, no matter how great the expense, surely this disease merited the effort. Success might mean a revolution in treatment for the illness. It would also mean, for its discoverers, extraordinary academic distinction, even possibly the holy grail, the Nobel Prize.

Many researchers in many institutions were to play a part in the hunt for the CF gene. However, three groups of scientists were to make most of the running.

The Hospital for Sick Children in Toronto lies on University Avenue. The avenue is a wide and generous boulevard that runs straight through the clean high-rise heart of the city to Lake Ontario. In that hospital in 1982 a post-doctoral scientist wanted to find the cause of CF. His name was Lap-Chee Tsui.

In Salt Lake City the Mormon religion had institutionalized the breeding of large families and the meticulous recording of family trees. Here Ray White had been amassing markers on the genetic map, in preparation for just such a search.

The molecular biology laboratories at St Mary's Hospital in London are housed in a structure built for the Victorian age and reached through a warren of shadowy corridors stuffed with dusty lockers and £50,000 centrifuges. At St Mary's, Bob Williamson and Kay Davies had already been mapping muscular dystrophy and had shown that RFLPs could locate disease genes.

In each country powerful charities and cystic fibrosis associations were dedicated to finding a cure for the disease, and were prepared to pour funding into any project that might achieve their aim. These charities also provided a network of people who would help to recruit the families necessary for finding the gene. The motivation of the association members was often their own families' suffering.

The biotechnology company, Collaborative Research, was the ingredient that caused the mixture to explode. Collaborative had spent ten million dollars competing for the genetic map with Ray White. That Collaborative hoped to make money was obvious. CF had the potential to make money because one in twenty people carries the disease gene. Screening the population or the relatives of affected children could dramatically reduce the number of babies born with the disease. A commercial test for CF would have an enormous market with a corresponding financial return. Collaborative's commercial ambition subsequently coloured the race for the CF gene in a highly unfortunate way.

The first problem faced by the investigators was to find sufficient chromosomal markers to map the gene. In contrast to muscular dystrophy, there were no chromosomal breaks to give a clue as to where the gene was hiding. In 1983 there were only 115 genes cloned, one-tenth of 1% of the 100,000 likely to be in the human genome. Of these, 32 were suitable for genetic mapping. There were 142 RFLPs. This gave a total of 174 map points to trace the CF gene. The best estimate at the time was that between 400 and 500 markers would be necessary to cover all chromosomes comprehensively. Finding CF was far from certain, and an enormous act of faith was needed to begin the research.

Ray White and his team were evolving strategies for mass production of markers. These were usually probes, pieces of cloned DNA that would stick uniquely to a specific place on a particular chromosome. Because the probes were in clones, which had to be grown in bacteria, it involved substantial time and money to distribute them to interested workers. (By contrast modern markers, such as microsatellite repeats, can be made by machines for DNA synthesis: all that has to be transported is the DNA sequence. The fax and electronic mail have replaced the parcelled tube of bacteria.) Ray White's team, despite their best efforts, were still a long way from a complete genetic map. Other probes could be gathered from various laboratories around the world. However, persuading someone who didn't know you from Adam to devote a day of work to preparing a probe for transport was not always easy.

In those days there was a feeling that probes might be valuable, or that they might lead to riches in ways that were not yet understood. This meant that the use of any probe usually involved signing a semi-legal document, commonly stipulating that the probe could not be used for commercial purposes without agreement, and that publication of any positive results should only be after consultation with the donating laboratory. Translated, it was a claim for part or all the money that might appear, and failing that, a claim for authorship on the relevant paper. In reality, spending a day preparing a probe for someone else is inconvenient, but it hardly warrants co-authorship on a paper. Similarly, in the absence of a patent, it would hardly lead to wealth generation.

It was in this atmosphere that the various groups set to work. By 1984 their combined efforts had covered less than 50% of the genome, without any sign of the CF gene. Collaborative were increasing their competition with Ray White to find new probes. They had generated about 150 markers, and at the end of the year they offered the probes to Lap-Chee Tsui. Quite naturally, he accepted. He began using their probes in August 1985.

Lap-Chee did not expect what happened next: almost immediately he found genetic linkage with a Collaborative probe. The lod score, which measures the strength of genetic association, was 2.8. This meant that the odds in favour of genetic linkage were 800 to 1. Although this probability was considered just short of proof, the finding was extraordinarily exciting. The probe was not in immediate proximity to the CF gene. Lod scores are calculated at a number of theoretical distances between a disease and a genetic marker. The point where the lod score is at a maximum is used to estimate the actual distance between the disease and the marker. In this case the statistics suggested the CF gene was within 15 million bases on either side of the Collaborative probe. It almost didn't seem to matter that no-one knew from where the probe came, that is on which chromosome it resided.

Finding
the cystic fibrosis
gene

The first hurdle on the way to establishing linkage had been overcome. It would only be a matter of time and effort before the gene was cloned and it was revealed how it caused cystic fibrosis.

Lap-Chee immediately contacted Collaborative and told them the news. They ordered him to keep it a secret, ostensibly while they worked out to which chromosome it belonged, and while they protected their investment with patents. This put Lap-Chee in an impossible position. He was unable to discuss his momentous findings with anyone else in the field, or to publish his results. He had competitors breathing down his neck. He was capable of finding which chromosome the probe was on without further inputs from Collaborative, although they had expressly forbidden him to do so.

His position had been made worse by a meeting of the CF researchers in Helsinki in August, held at the same time as the Toronto lab were first finding linkage. An 'exclusion map' was drawn up, which combined the negative results of all workers in the field. Markers on most chromosomes showed clear evidence of non-linkage. However, three chromosomes were almost uncharted and CF was likely to be on one of these three: chromosomes 8, 18 and 7. Lap-Chee tried to get information from Collaborative about the localization of the probe but if they knew they weren't telling, and said they still wanted time to sort it out.

Finally Lap-Chee ran the tests himself. The probe, and the CF gene, were on chromosome 7. He told Collaborative Research, and, according to *Science* magazine's account of events, Collaborative were greatly displeased. Nevertheless, patents were submitted by the end of September, simultaneously with a paper to *Science*. The paper described the linkage, but did not say where the linkage was.

By now the rumours were flying in the relatively small circle of gene mappers. At an October meeting of CF researchers, Lap-Chee presented his linkage results. By now the lod score was 4, which made linkage certain, but Collaborative would still not allow him to say that the gene was on chromosome 7.

After the meeting there was an enormous burst of work from Williamson's and Ray White's teams, as both tested every chromosome 7 marker they had. There were a number of possible reasons why they had simultaneously decided to go for broke on chromosome 7. One possibility was the clue given by the exclusion map presented in Helsinki. Another was that there had been a leak of information from the Toronto group. If there had been a leak, should White and Williamson have reacted so strongly to a rumour of someone else's result? Given the huge level of commitment of these big groups to the hunt, I think it was almost impossible for them not to act on whatever information came their way. The chromosome 7 markers were positive in families on both sides of the Atlantic. Two papers were submitted to *Nature* and Collaborative got wind of them. They insisted that *Nature* accept a paper from Toronto and publish it with the other two. They may also have influenced an editorial that accompanied the three papers when they appeared together in November. The editorial specifically mentioned rumour as a source of the chromosome 7 linkage.

By now everyone was pretty upset. At the time it was fashionable to cast aspersions on White and Williamson. In an ideal world they might have discussed their results with Lap-Chee. However, Collaborative were suppressing findings that, under the rules of normal scientific behaviour, should already have been made public. In these circumstances it could be argued that it would be very difficult to pick up the phone and ring Lap-Chee for a chat.

Much of the blame must lie with Collaborative. It is not only that they were acting for commercial gain. Although they claimed their behaviour was standard commercial practice, the linkage result depended mostly on the charities who had funded the Toronto research and the efforts of the families from the CF associations. Nevertheless, a widely publicized statement from Collaborative that 'we own chromosome 7' caused the company's stock to rise sharply, although the attempt to patent chromosome 7 was always likely to fail.

The fighting and the name-calling obscured the real scientific progress that the three papers announced. Not only had the gene been localized on chromosome 7, but one of Ray White's markers, called *met*, was within one million base pairs of the CF gene. Williamson too had a close marker, called J3.11.

The next step in the study was to discover where the new markers were in respect to the CF gene. J3.11 could be on the same side of CF as *met*, or it could be on the other side. If it was on the other side then the two markers would 'flank' CF, defining an interval that must contain the gene. To decide if *met* and J3.11 flanked CF required more families than any single group could muster. All the interested scientists agreed to pool their data on the chromosome 7 markers. The results showed the value of open collaboration: *met* and J3.11 did indeed flank CF.

This finding made possible for the first time the antenatal diagnosis of cystic fibrosis. If a family already had a child with CF, then the new markers could detect if a sibling foetus had CF. The parents and the affected child had to give a sample of blood to be typed for the markers, and a sample of placenta or amniotic fluid had to be taken to test the foetus. Typing meant looking for RFLPs in the parents and children. The RFLPs allowed the laboratory to see which parent was giving which chromosome 7 (or which relevant region of chromosome 7) to the children. As an example, a father might have, in his two chromosome 7s, RFLP types 'A' and 'B', and the mother might have 'C' and 'D'. If their affected child had RFLP types 'A' and 'D'('A' from the father, and 'D' from the mother), and had the disease, then, as CF is recessive, and requires two abnormal genes, the 'A' and 'D' chromosomes must both carry a diseased gene. If the foetus on which the antenatal diagnosis was being performed had type 'B' and 'C', then it had not inherited either abnormal chromosome, and the parents could be reassured accordingly. If, on the other hand, the foetus had 'A' and 'D', like its affected sibling, then it too would be likely to develop the disease. In some families the procedure found with 99% accuracy if the foetus was affected. However, in many other families the risk was much less sure: if one parent was 'B''B', for example, then the abnormal chromosome could not be differentiated. As more and more became known about the region around the CF gene, and eventually about the gene itself, the easier it became to be sure which chromosome 7 was which and the more precise antenatal diagnosis was to become.

That *met* and J3.11 flanked the CF gene made its eventual discovery inevitable. They were one million base pairs apart. One million bases is the upper limit of a piece of DNA that can be 'physically mapped'. Physical mapping means the actual dissection of the DNA in a part of a chromosome, ending with the sequence of the DNA and the identification of its genes. 'Genetic mapping', with markers and lod scores, only gives a statistical probability that a gene is near to a particular point. Physical mapping is preferable to genetic mapping because it gives concrete information about genes: you know where you are with a physical map. However, it requires an enormous amount of work. Any more than a million bases would have made physical mapping impossible, at least in 1985.

The next success in the race was Williamson's. His group decided to use a mouse cell line containing a small piece of human chromosome 7. A 'cell line' is a culture of cells that are descendants or clones of a single original cell. The line contained fragments of human chromosomes in mouse cells. Even though these hybrid cells contained mixed mouse and human DNA, human DNA could be recognized by its unique repeated sequences.

The cell line that Williamson studied contained *met*. Because *met* is an oncogene, or cancer gene, it encourages cells to grow. Mouse cells with human DNA containing *met* could be identified by their abnormal growth. If they also contained J3.11, then they were also likely to have the human CF gene lying between *met* and J3.11.

This approach was technically very difficult. Williamson knew, however, that if the hybrid experiment worked, the CF gene and a small amount of flanking DNA could be obtained in quite a pure form. The alternative was to 'walk' between J3.11 and *met* by breaking chromosome 7 into minute pieces, and cloning the pieces into bacterial plasmids.

A chromosome walk starts with a single cloned fragment of DNA, for example a clone containing J3.11. The clone is used to find another clone that overlaps, so that a step is taken sideways along the chromosome. In turn the new clone identifies yet another overlap, and so on, gradually extending the walk until a flanking marker such as *met* is reached. At the end of a walk all the DNA between the two markers has been identified, and is held in clones. Thus DNA has to be cloned, and the clones painstakingly ordered, before it can be examined for the presence of genes.

In 1985, the longest piece of DNA that could be cloned into a plasmid was 2,000 base pairs, and most clones contained much less than this. Each step in a walk took one person about a month. Because of overlaps, this step would, on average, cover about 500 base pairs. To walk 1,000,000 base pairs would take one researcher 2,000 months, an impossibility. With improved cloning 40,000 bases of DNA could be inserted into bacteria, reducing a million base pair walk to 100 person-months; still a very long time.

It was therefore gratifying when the *met* hybrid approach worked. Williamson could hold in a test tube the million bases of DNA that must have contained the CF gene. Even within that region there might be many genes, and there would certainly be hundreds of thousands of bases of meaningless DNA. Sequencing a million base pairs was impossible, so to detect the genes Williamson took another gamble. He decided to concentrate on 'HTF islands'. HTF stands for 'Hpa tiny fragments'. Hpa is another restriction enzyme that cuts DNA. The regions of DNA that are cut into tiny fragments with Hpa are often found at the beginning of genes, though not all genes have HTF islands and why they sometimes mark genes is still unknown. At the time it was simply a phenomenon observed by an Edinburgh geneticist, Adrian Bird.

Williamson and his group found a gene near an HTF island in the *met*–J3.11 interval at the beginning of 1987. The gene was expressed in the lung and it was in 'linkage disequilibrium' with CF, which means that a particular polymorphism in the DNA around the gene was found more often than expected in children with CF. Linkage disequilibrium between a gene and a disease is only seen if the gene, or another gene extraordinarily close to it, causes the disease.

The St Mary's group were sure that the CF gene was theirs, and in a paper in *Nature* in April they announced the discovery of their 'candidate'. Again there was a huge amount of accompanying publicity but final proof that the gene caused CF was still lacking: the gene had not been sequenced and no mutation had yet been found in children with CF.

During the next months it was increasingly clear that the candidate was not the real CF gene: information from new families showed it did not map in precisely the right position; and when its sequence became available, the gene did not contain any mutations. Williamson was wrong. However, he had found that the failed candidate, known as IRP, must be closer to the real gene than any other marker. He had also shown that HTF islands could pinpoint genes within a stretch of DNA.

Again a war of words broke out. Williamson released IRP freely to diagnostic laboratories for the antenatal diagnosis of CF, but refused to let rival groups use the IRP gene to help clone the CF gene. Other investigators, including Ray White, complained that Williamson had put them off the track. They argued that they had believed Williamson and given up their research efforts, and that they could not now make up their lost time. It seems to me that to have given up so easily suggests a lack of commitment, and that a reason for it was that Ray White was becoming increasingly more interested in a general map of the genome; by contrast Lap-Chee in Toronto was still working flat out.

Lap-Chee had decided to push on with genetic mapping to find new markers between *met* and J3.11. He and his team tested over 200 before they found two new markers. It had originally required nearly 200 probes, tested in an international effort, just to locate the gene on chromosome 7. The 200 new probes therefore represented a formidable effort by the Toronto team.

Of the two new markers, one was close by Williamson's IRP gene, and was likely to be within a quarter of a million bases of the CF gene. This distance was walkable, just. Lap-Chee formed a collaboration with Francis Collins, a gene hunter who now leads the American Human Genome Initiative. Collins had invented a method of 'gene jumping' which enabled much larger steps to be taken in the walk along the chromosome 7 DNA. Gene jumping was technically very difficult, involving the use of delicate persuasion to form the DNA into enormous loops. It has not subsequently been a popular technique, but the collaboration speeded the process of searching along the chromosome from IRP to the presumed site of the CF gene itself.

In January 1989 the rumour in a very small, informed circle of geneticists was that Lap-Chee had cloned the gene. By the time of the international gene mapping meeting at Yale in early summer, the rumour had become much more general. The rumour was fuelled at the meeting because Collins had cancelled a planned appearance without explanation. It was assumed, rightly or wrongly, that this could only be because he had a secret he wanted to keep. Lap-Chee was booked to give a plenary lecture on CF, which was scheduled so that everyone at the conference could attend. This he gave, talking the audience through his evidence for closer and closer localization of the gene. It was a complicated lecture, and many of the audience were nursing conference hangovers, but at one point it was obvious that if he were to make an announcement, then this was the moment. You could have heard a pin drop. He said nothing new.

The next morning Lap-Chee was chairman of the chromosome 7 meeting. Representatives of all his principal competitors were in the audience. He opened with the remark, 'If anyone has cloned the CF gene could they please come forward and tell us now.' No-one stepped up.

At the beginning of September, in three papers in the journal *Science*, Lap-Chee and his collaborators triumphantly reported the cloning of the CF gene and the discovery of a mutation, delta-508, that caused the illness. These papers described very substantial pieces of science, at least a year's work. Lap-Chee knew he had the gene in New Haven, but no-one guessed.

In fact, Francis Collins had indeed visited the New Haven conference, meeting Lap-Chee late at night in his room. Together they went through the results that had been faxed from their laboratories. Lap-Chee had discovered the delta-508 mutation earlier in the year, but at that stage he had no way of telling if the mutation was likely to cause disease, or if it was a harmless variant of normal. The faxes coming in to Collins and Lap-Chee showed conclusively that the mutation was only found in children with CF. For the first time they were sure that the race for the CF gene was over. Lap-Chee, universally recognized as the nicest of men, had won as he deserved. Williamson, who had contributed so greatly, was gracious in defeat.

Much had been made of the clash of personalities in scientific circles and in the scientific press during the five years of the race. A quotation from *Science* magazine in 1988, 'This is not your average ego-driven science: this is nasty!' summed up the attitudes of the race watchers. This view of things should not obscure the extraordinary scientific progress made during those five years. Without Bob Williamson and Ray White the search for CF would have taken much longer. The scientists involved were hard men working in the forefront of a new science on an extraordinarily difficult problem. Normal well-balanced people do not work that hard or care about success that much, and their mistakes are not blown out of proportion by a sensation-hungry press. Outstanding sportsmen are rarely articulate gentlemen; it is perhaps unfair to expect a real scientist always to behave impeccably. What goes into the textbooks are the results. Above all, it was Collaborative Research's zealous commercial instincts that set the tone for all the acrimony.

Partly because of the wars over the CF gene, genetic journals now only accept papers if the relevant sequence has been given to a public database. Probes and genetic libraries are similarly stored in central repositories so that all can have access to them.

Finding the CF gene has not produced a dramatic cure for the condition, but it has produced an explosion of understanding. Lap-Chee had discovered that the gene codes for a big complex protein that sits in the surface membrane of cells in the lung and sweat glands. The protein acts as a channel for the salt and water pushed into the body's secretions. The gene is long, 250,000 base pairs, about three times as big as an average gene.

In all this sequence only three bases are missing in two-thirds of the children with the illness. For the CF gene hunters to have identified these three base pairs from the three billion that make up the human genome is the equivalent of finding three particular people among the population of the whole world.

These three bases code for one amino acid, phenylalanine. The absence of this single molecule from the 1480 amino acids that make up the protein is sufficient for it to fail in its function. Around a hundred other CF mutations have now been found. Some of these affect the gene's function less than others, causing more mild cystic fibrosis, or cystic fibrosis with marked pancreatic disease, or even a variant of the disease so mild that it only causes illness in middle-aged smokers.

CF is a recessive illness because one functioning gene is more than enough to make normal secretions. Indeed, only a few percent of normal function is enough to prevent disease. This has important implications for treatment of the illness: even a partially effective therapy may be enough.

Finding the CF gene has not yet produced a cure. Nevertheless, molecular biology has come up with one new and dramatic treatment for CF which is a terrific example of science at its best.

Steve Shak was a respiratory physician who now works for Genentech in California. He became interested in treating the sticky mucus of children with the illness. Mucus that is infected and coughed up is called sputum. Many adults with lungs damaged by other illnesses also cough up thick tenacious sputum, and as part of an examination of the lungs a physician will inspect the sputum for colour and consistency. The more infected the sputum is, the thicker it is (and the more colourful!). Infected sputum is very viscous, and pours like a raw egg, which is why children with CF find it very difficult to clear the sputum from their lungs and require hours of physiotherapy every day.

Shak asked himself a very simple question: why is the sputum in CF sufferers so thick? The question was so simple that no-one else was asking it. Like a good scientist he went to the scientific literature and found that, years before, someone had published a biochemical examination of sputum.

To Shak's surprise he found that sputum contains a very high concentration of DNA. The DNA has come from white blood cells which attacked infection in the respiratory mucus. In the process they died, eventually dissolving and releasing their DNA into the mucus.

The same kind of release occurs when a molecular biologist purifies DNA for his experiments. He or she starts with cells suspended in a solution. The solution pours easily, just like water. A strong detergent is added to the liquid, to break the cell membranes and release the DNA. The detergent immediately destroys the cells, and instantly the liquid becomes viscous and 'gloopy'. This is because the very long DNA molecules have suddenly unwound into the solution.

Shak reasoned that if he could break up the DNA in some way, then he might help sputum to clear from the lungs more easily. He knew of the existence of an enzyme called DNAse, which had been extracted from the pancreases of cows and pigs and which destroys DNA. In the past it had even been tried as a treatment for lung diseases. However,

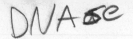

because it was a non-human protein, allergic reactions had prevented its use. Shak searched to see if the human DNAse had been cloned. It had not, and so he quickly moved to clone it, to capture its DNA sequence.

After he had cloned the DNAse gene he put it into a cell that produced protein from the gene. The protein could then be recovered from a broth in which the recombinant cells grow. Shak took the recombinant human DNAse and added it to a test tube half-full of sputum. As a control he had a second test tube, to which he did not add DNAse. After half an hour he inverted the tubes. The sputum treated with DNAse flowed freely down the side of the inverted tube, while the untreated sputum remained in a sticky blob. The next thing Shak did was to patent the human DNAse gene.

Genentech moved very quickly to produce large amounts of highly purified DNAse, and to try it out on children with CF. It worked, rapidly improving the lung capacity of the children. Today, there is a big new building at Genentech devoted to the manufacture of DNAse.

Steve Shak's story shows that science can advance through the work of single individuals with a unique insight. However, for Huntington's disease, the last great single gene disorder, it was necessary for hundreds of scientists in many laboratories in many countries to cooperate.

CHAPTER 6

THE HORROR, THE HORROR

Huntington's disease is named after the good Doctor George Huntington, a physician on Long Island in the nineteenth century. His father and grandfather were also doctors. They had seen the illness in their practice before young George brought the attention of the medical world to its presence.

Huntington's disease is an awful and dramatic illness that comes on in middle age. Sufferers develop chorea, sudden flinging movements of the limbs that are quite involuntary and uncontrollable. George Huntington saw his first cases of hereditary chorea when he was eight:

> Driving with my father through a wooded road leading from East Hampton to Amagansett we suddenly came upon two women, mother and daughter, both tall, thin, almost cadaverous, both bowing, twisting, grimacing.... From this point on, my interest in the disease has never wholly ceased.

Years later he was to write his description of the illness, which he called 'the hereditary chorea'. His description is wonderfully clear:

> Its most marked and characteristic feature is a clonic [alternating between contraction and relaxation] spasm affecting the voluntary muscles. There is no loss of sense or of volition attending these contractions, as there is in epilepsy; the will is there, but its power to perform is deficient, the desired movements are after a manner performed, but there seems to exist some hidden power, something that is playing tricks, as it were, upon the will, and in a measure thwarting and perverting its design; and after the will has ceased to exert its power in any given direction, taking things into its own hands, and

keeping the poor victim in a continual jigger as long as he remains awake, generally, though not always, granting a respite during sleep.

One has to feel the spirit of chorea dancing and jerking through the 108 words of the second sentence; Huntington's language conveys much more of the essence of the disease than a dry statement of facts.

There are other causes of chorea. In the middle ages an extraordinary outbreak of 'dancing mania' affected the populations of northern France, the Netherlands and Belgium. Victims, according to Hecker (1888),

> ...formed circles hand-in-hand, appearing to have lost all control of their senses and continued dancing regardless of the bystanders for hours together in wild delirium, until at length they fell to the ground in a state of exhaustion...

The disease spread through Europe over a generation. It was given the name of 'St Vitus' Dance' after a fourth-century Sicilian, Vitus. A similar condition, called 'tarantism', after the spider, was very common in Sicily. Vitus was martyred for his Christianity in a cauldron of boiling lead and pitch. He is said to have prayed for the salvation of all those who suffered from the dancing plague, and that his prayer was personally answered by the Lord as he went to his super-heated death.

The dancing plague was almost certainly a phenomenon of mass hysteria. Even in modern society, newspapers still occasionally report outbreaks of equally bizarre illnesses that have suddenly spread through a school or a village.

Thomas Sydenham applied the name 'St Vitus' Dance' to another convulsive disorder of movement that he had seen in children. We now know this syndrome to follow sore throats caused by the streptococcus bacteria. The victims usually recover completely. Over the years the term 'St Vitus' Dance' has been gradually replaced by the less picturesque name of 'Sydenham's chorea'. This is a shame for lovers of language, but perhaps a relief for parents whose children have the illness. Chorea comes from the same word as the Greek for chorus: a group of dancers and singers.

Huntington knew of the existence of Syndenham's chorea, and so he explained how the illness he described differed from common chorea:

> ...there are three marked peculiarities in this disease...its hereditary

nature;...a tendency to insanity and suicide;...its manifesting itself as a grave disease only in adult life.

The bizarre movements of chorea become inexorably worse, and are followed by a general decline in mental function. Eventually dementia sets in. Death, which follows ten or twenty years after the first inexplicable twitch, comes as a merciful release. Because the condition is so distinctive, it has always been obvious that it was hereditary. Although the illness does not appear till late in life, almost everyone with the abnormal gene is doomed to develop the condition. The late onset of the disease means that children of a sufferer will have been born by the time of their parent's diagnosis. Thus, although the abnormal gene is fatal, there has been no check on its spread through the world's population.

The study of families with the illness began long before any understanding of genes or genetics. However, as Michael Hayden observes in his monograph *Huntington's Chorea*, tracing the family history of the disease can be more difficult than with other illnesses. This is because members of affected families are likely to deny hereditary madness. Nevertheless, it has been possible to draw up pedigrees of Huntington's disease right back to the seventeenth century.

The earliest affected persons in American are known by the pseudonyms of Jeffers, Nicolas and Wilkie. All three had been born in the English village of Bures, near Colchester on the Suffolk–Essex border. They had sailed from Great Yarmouth to arrive in Salem in 1630 and it seems that they were difficult men, because records exist showing each of them falling into trouble with the law. There have been other sources of Huntington's disease in America, and throughout the rest of the world. However, from the genealogical evidence it seems likely that the most widespread mutation in the gene arose for the first time somewhere in north-western Europe, in France, Holland or Germany.

Hayden has observed that, in formerly colonial territories, persons with the abnormal gene were often found among the earliest pioneering families. This seems to have been the case in America, the West Indies, South Africa and Australia. Their original communities may have rejected these individuals, perhaps because of sociopathic personality changes in the very early stages of the disease. Hayden, however, suggests that the emigration of so many men and women with the Huntington's gene was an attempt to escape their families and their fate. The ancient Greeks would have known better, that destiny cannot be denied. Even in the New World the lives of Jeffers,

Nicolas and Wilkie ended in madness and despair. Their descendants fared no better, for they included no less than seven of the females to be regarded as witches in an age when witchcraft was punishable by burning at the stake.

Huntington's disease is thus a severe and horrible illness, dramatically genetic, which had been internationally recognized and studied before the era of DNA. It is a 'dominant' condition, dominant because it is handed down in an unbroken line running through the generations. A dominant inheritance means only one copy of the abnormal gene is necessary for the illness to occur. Fortunately Huntington's disease is rare, affecting one in 20,000 people; nevertheless, it is easy to understand why it became a prime target for the first gene hunters.

The hunt began in the laboratory of James Gusella, at the fabled Massachusetts General Hospital in Boston. There had been many scientists working cooperatively on Huntington's disease since 1970 and in 1980 Gusella and his colleagues began looking for families with Huntington's disease whom they could use to find the gene. They chose two large families, one from the United States, and one from the remote shores of Lake Maracaibo in Venezuela. The Venezuelan family descended from a single common ancestor with the illness, in a pedigree containing over a thousand people. It took Gusella and Nancy Wexler and their Venezuelan colleagues three years to work through the ramifications of the family. In a labour of love, the Americans visited for a month each year, examining everyone meticulously, and collecting blood samples for later DNA analysis.

At the beginning of 1983 they decided to go hunting. Initially they used non-DNA markers, that is they studied the slight variations in some proteins found in the blood. These variations can be used for genetic linkage in the same way as DNA variants, because the protein is made from DNA (a gene) with a particular chromosomal location. There are about fifteen or twenty proteins that can be used in this way. Unfortunately, the information they provide is limited because they do not, on average, differ enough between individuals and the protein polymorphisms excluded Huntington's disease from only about 10% of the genome. Gusella and his team then decided to try DNA markers, an approach that was at the time only possible in theory.

Ray White and his probe factory had not yet produced a map of the genome, so Gusella and his team had to find their own markers. They asked a leading molecular biologist, Tom Maniatis, for a library of human DNA clones. The library held millions of clones, which could be grown in bacteria, each holding a very small and specific part of the human genome. This sort of library is now commonplace, but at the time it was a precious resource.

In an initial experiment Gusella and his team found twelve clones that detected polymorphisms with which they could try to map the gene. One of these, the eighth clone (called G8), showed tight genetic linkage to Huntington's disease: G8 and the Huntington's disease gene must be close to each other on the same chromosome.

This was an absolutely outrageous piece of luck. They were the first group in the world to use DNA markers to look for genes and, against odds of at least 200 to one, they had found linkage with their first batch of markers. It was only a minor drawback that they had no idea from which chromosome G8 had originated, for they were able to show quite quickly that G8 was on chromosome 4.

This had two very important implications. First, in families known to contain affected members, the disease might be diagnosed before symptoms had developed. Affected individuals could then avoid pregnancy, or if pregnant they could submit their foetus to antenatal testing. It came as a surprise, though perhaps it should not have, that not everyone who was potentially affected with such an awful illness would want to know their fate while they were still hale and hearty. The second great promise offered by linkage to chromosome 4 was that the gene itself would be cloned, the cause of the disease made clear, and effective treatment invented.

After such a bright start, rapid progress seemed inevitable. An editorial in *Nature* talked of 'The beginning of the end'. Sadly, after that first great stroke of good fortune, the luck was to run out for Gusella and everyone else who joined in the search for the gene.

The researchers began to look for new markers on chromosome 4. Quite quickly they found that the gene was very near the tip of the short arm of the chromosome (chromosomes usually have one short and one long arm). The tips of a chromosome, called the telomeres, are very complex compared to other chromosomal parts, containing very many repeated sequences. This structure is necessary for the faithful copying of chromosomes before cell division. The region through which the scientists had to search for the Huntington's gene contained about six million base pairs of DNA.

In order to map the gene within the six million base pairs the researchers looked for 'recombination events'. These are places where chromosome pairs have broken, crossed over and rejoined. In contrast to the sort of chromosomal breaks that cause Duchenne muscular dystrophy, recombination is a normal event. It is part of the shuffling process that mixes genes into new packets for each generation. There is more about crossovers in the chapter on sex (see page 85 ff.). Although the positions of these crossovers are not visible, they can be found with genetic markers, and reveal a point between the normal chromosome and the region that holds a disease gene: inheritance of the DNA on one side of the crossover may give a normal child, but inheritance of the other side causes disease. The mutant gene is thus localized. Within six million base pairs, recombination events are quite rare, and will be found in about six of every hundred children.

Within a year the scientists had found twelve recombination events. Unfortunately they gave conflicting evidence. Some seemed to exclude Huntington's disease completely from the entire region; others placed it near G8, about two million base pairs from the tip; whereas still others suggested it was right at the end of the chromosome, within a few hundred thousand base pairs of the tip. The data seemed to make most sense if the gene was within this terminal region. The inconsistencies were put down to second recombinations that had not yet been detected. Since recombinations on a chromosome segment are very rare, with odds of a thousand to one even if the segment is as big as six million base pairs, this explanation was unsatisfactory. However, it was the best available, and most of the researchers pushed on, accepting the hypothesis.

The subsequent slow progress of the Huntington's search may have been because of this uncertainty. It took until 1990 for the terminal region of the chromosome to be cloned, in the lab of Hans Lehrach at the Imperial Cancer Research Foundation in London. Sadly, there was neither a second recombination event, nor the fabled Huntington's gene. There was only a mess of scrambled, repetitive DNA. The hunters had found a junkyard instead of a gold mine. They had gone to the wrong place.

The number of scientists working on the problem grew steadily on both sides of the Atlantic. Most leading groups joined, under the collective name of 'The Huntington's Disease Collaborative Research Group'. The group's name may not have been particularly imaginative, but collaboration was vital if there was to be a successful outcome to the gene hunt. Some workers scoured medical clinics and genetic registers for new families with the illness, in search of new recombination events that would help resolve the localization of the gene; others attempted to find new markers; others began to piece together the 'physical map' of the DNA. They were looking for landmarks, such as HTF islands, which showed up the potential sites of genes, rather as a geological survey might look for signs of ore-bearing rock.

By this point it had become impossible to ignore the fact that the inheritance of the disease was not at all straightforward. It was noticed that the illness seemed to get worse as it passed down through the generations. Those whose disease came on later in life had children in which the worsening was not very marked; however, if someone was affected earlier in life, which usually meant the disease was more severe, then the disease would be much worse in the next generation.

Even in 1985, the laws of genetics and the understanding of genetic mechanisms did not allow for traits to alter with passage through generations: except for episodes of distinct and sudden mutations, the gene was held to be fixed. This very puzzling finding was further compounded by the realization that particularly severe cases were usually transmitted through the mother. Again there was no genetic explanation, although genomic imprinting, discussed on pages 103–105, was half-heartedly suggested as a possible cause. The way in which the issues were dealt with shows how human failings affect scientists as much as other mortals.

The worsening of some genetic illnesses through the generations was first suggested by a Doctor F.W. Mott at the beginning of the century. He developed his ideas from the study of families with mental illness, or in the jargon of the time, lunatics in madhouses. He observed, or thought he observed, that the children of the mentally ill were even more seriously affected than their parents. This he called 'anticipation'. His conclusions were highly coloured by the eugenic view propounded by Francis Galton (about whom more later). Mott maintained that society was degenerating, because,

> At the present time in Great Britain restriction of families is occurring in one half or two-thirds of the people, including nearly all the best, while children are to be freely born to the feeble-minded, to the pauper, to the alien Jew, to the Irish Roman Catholic, to the thriftless capital labourers, to the criminals and others.

To Mott the insane were like 'rotten twigs dropping off the tree of life'. That he could proffer these views as serious science in the *Lancet* and the *British Medical Journal* in 1910 tells us rather more about British society then than we might wish to acknowledge. Mercifully, such opinions declined in popularity, and F.W. Mott, too, dropped off the tree of life.

Genetic anticipation was then recognized, although under another name, in families with the illness of myotonic dystrophy. This is a rare inherited muscular disease whose victims are unable to relax their muscles after contraction. Sufferers have difficulty walking, and men develop premature baldness, and so have a fairly typical appearance. When I was a medical registrar I worked for the great A.K. Cohen, President of the Royal Australasian College of Physicians. Cohen's powers of observation were legendary and he caught out his registrars and housemen with distressing regularity. One night, in casualty, I diagnosed a case of myotonic dystrophy. The man had been through several doctors' surgeries with the illness quite unrecognized. Full of pride at my cleverness I prepared to trap the great man the next morning. As Cohen's ward round started, the patient shuffled between his room and the lavatory, some forty yards away from us at the end of the corridor. 'Dystrophia Myotonica,' said Cohen, barely looking up from his papers, 'is that the interesting case you wanted to show me?'

To this there was no reply. Unfortunately for those with myotonic dystrophy, diagnosis is often delayed, as most medical practitioners lack the observational skills of Alex Cohen. Cohen was a regular censor at specialist medical exams, and tells the story of an examinee who, as physicians are taught to do, shook hands with the patient whose dystrophy she was meant to diagnose. Cohen and the other examiners watched as the examinee disentangled herself from the patient's grip. When asked to test his handshake again, she again disentangled herself without suspecting the diagnosis. Her inability to recognize the obvious was perhaps because of nerves, but might also have resulted from a prejudice which assumed that even ordinary old men tended to grasp at young female hands and hold them too firmly.

A clinging grip and baldness aside, people with myotonic dystrophy also suffer from cataracts. In 1918 the Swiss ophthalmologist, Bruno Fleischer, recognized that the cataracts were often present in the parents or grandparents of patients with myotonic dystrophy, despite the fact that these individuals might show no other evidence of the illness. The phenomenon was consistently observed and reported with gathering scientific authority until the late 1940s.

The problem then became the increasing knowledge of the physical basis for genes and genetics. Within the contemporary understanding of how genes worked there was no place for curiosities that led to a generalized change in a gene in many families over a limited number of successive generations. Genetic anticipation could not be real, so the earlier observers had to be wrong. Sure enough, Lionel Penrose, a gentle man and a superb geneticist, carried out an elegant analysis. He showed the appearance of genetic anticipation was all down to bias in the way the families had presented to the doctors. Simply put, he concluded that only the worst cases would attend a medical practitioner, so that the cases who did not go to the doctor, who would most likely be parents, would be mild by comparison.

Relieved of the responsibility of trying to understand the 'impossible', the medical and scientific fraternity then largely forgot genetic anticipation. All was well until another genetic illness came to attention. This condition is known as fragile X syndrome. Children and adults with the condition are of low intelligence, with wide-set eyes. Males with the disease have large testicles, which everyone concerned with the illness seems to remember. The pattern of inheritance is sex-linked, like muscular dystrophy, although women can be affected to a greater or lesser degree and can inherit the condition from their fathers. This meant that a gene on the X chromosome was likely to be causing the illness. Sure enough, under the microscope the X chromosome could even be seen to be abnormal in shape, with a small piece hanging off one end. Fragile X syndrome also seemed to get worse through the generations, and to make matters worse the most severe cases were always inherited from the mothers' side.

Peter Harper is a Cardiff geneticist who played a leading role in the hunt for the genes for myotonic dystrophy and Huntington's disease. He was the first to point out the similarity between the anticipation seen in fragile X and in myotonic dystrophy. This very clever observation passed largely unnoticed, because it still could not be explained. Again when I was a medical registrar, an elderly man was admitted to the ward one night; he had suffered a mild stroke that had damaged only a small part of his brain, and the stroke had affected his ability to follow objects with his eyes. A neurologist calls this sort of eye movement 'pursuit'. Now, although the man could not follow a finger passed before his gaze, he could, on command, flick his eyes from side to side quite normally. He was able to do this because the two kinds of eye movement, following and flicking, are controlled in two separate areas of the brain, and the old man's stroke had only damaged one of those areas.

An able neurologist who saw the patient the next day made the diagnosis. Because the poor old man with his stroke was an 'interesting case', a succession of medical students came to see him. They were asked to examine his eyes without being told of the diagnosis, and they all reached the same conclusion: the patient was demented. He could not even understand when he was asked to follow a student's pen with his eyes, although he would easily turn his gaze to look when a nurse came into the room. Almost without exception the students became irritated by this 'stupidity'.

There is a moral to these medical stories and to the history of genetic anticipation. It is that we are all to a greater or lesser degree like those students. Faced with something we do not understand we dismiss it as artefact, because this is the easiest way of dealing with puzzling problems. The art of science, or of being a good physician, is to seize on the incongruous and pursue its explanation.

The riddle of genetic anticipation was solved in a dramatic way when the fragile X site was cloned in 1991. The gene, christened Fra-X, contained a sequence of DNA with three nucleotide bases that read CGG. This 'trinucleotide' was repeated many times, both in the normal and the abnormal sequence. However, the number of repeats differed greatly between normal and diseased genes. In normal people the CGG triplet is repeated ten to thirty times; in people with fragile X syndrome the number of repeats is much greater, between fifty and two thousand. This floppy piece of DNA interferes with the normal function of the gene. The more repeats there are then, naturally, the more severe the illness.

Carriers of the illness had between forty and fifty repeats, not enough to stop the gene working; but this number of repeats was unstable and during cell division, when an ovum was made, the number of repeats could suddenly double or triple. Genetic anticipation was at last explained, and was enthusiastically welcomed into the genetic community.

By the end of the same year the gene for myotonic dystrophy had been found. Sure enough, it also contained a trinucleotide repeat that destabilized the gene and caused illness. Again normal people had 20 to 35 repeats, carriers had more than fifty, and people with the disease had between a hundred and two thousand. Within a few months a third illness, much rarer than the others, called spinobulbar muscular atrophy, was also shown to be due to expansion of a repeated sequence.

Attention now turned back to Huntington's disease. The presence of genetic anticipation might suggest the mutation's appearance, but the problem of localization still made further progress seem hopeless: the gene was somewhere in the middle of six million base pairs of DNA, and there were perhaps another three hundred genes in the same region. Even with the very tip of the chromosome now excluded the genetic information from the position of the recombination events

New Type of Mapping

was hopelessly ambiguous.

Instead of giving up, or perhaps because they were now in so deep that they could not give up, the scientists turned to a different means of mapping. This was based on a process called linkage disequilibrium. Most cases of Huntington's disease, it was believed, were descended from a single ancestor. The mutation would have occurred on one of his chromosome 4s. That chromosome would have had the usual number of polymorphisms around the Huntington's gene, and these polymorphisms would have been common in the general population. On its own, a single point variation near the Huntington's gene would not help gene hunters twenty generations on. However, if all the polymorphisms in the region were considered, they would give a highly individual signature. It was possible that the signal might still be detectable after hundreds of years, even though the signature would eventually be lost, as all the DNA in the region was shuffled and re-shuffled though succeeding generations. In the jargon, the polymorphism would then be in equilibrium in the population. The signature would first begin to fade far away from the mutation in the Huntington's gene, and would persist longer close to the gene. Where it persisted there would be disequilibrium.

Gusella and the Huntington's Disease Collaborative Research Group examined their families for the presence of disequilibrium. They found a faint signal, about two million base pairs from the tip of the chromosome. The signal showed that one-third of all cases shared the same part of the chromosome. They were descendants of the same person, the first to have Huntington's disease. The signal was not definite, and the probability that it was just a chance finding was high enough to convince most people to keep looking elsewhere. Gusella and his team took the risk and concentrated on the region, which was about 500,000 base pairs long.

They cloned the entire area. Half a million base pairs are too long to sequence, and so they had to use an exotic piece of molecular biology called 'exon amplification' to trap pieces of the genes from the region. They found three genes, none of which contained any abnormalities; then, towards the end of the 500,000 base pairs, they found another gene, which they called IT15. The gene was large, spanning 200,000 base pairs. It made a completely unknown protein, which had no similarities to any other known proteins.

In the gene they found a sequence where the code CAG was repeated many times. In the normal population the repeat had between 11 and 34 copies; in the patients with Huntington's disease the number of repeats was between 42 and 66 copies. The longer the number of repeats, the worse the illness. This was exactly like the triplet repeat systems already found in fragile X syndrome and myotonic dystrophy. The hunt was over. The gene for Huntington's disease had been found. It had taken over ten years, in a battle involving hundreds of researchers in different laboratories and costing millions of pounds.

Though there was general jubilation at the end of the search, finding the gene does not mean that there will be an instantaneous cure for the disease. Many new genes have recognizable motifs that give a hint as to their possible function. This was not so with the 'Huntingtin' gene, as it was christened. Its role is as yet unknown. Mysteriously, it is active in many tissues, but only causes disease in the brain. Another decade of work awaits, but Gusella and all the others who have taken part in the hunt have been freed from their Sysiphean task.

So the search for the Huntington's disease gene was characterized by an amazing initial stroke of luck and was then beset, and almost overcome, by difficulties. Other researchers have been cursed with the same luck as Gusella.

Steven Reeders, formerly of the Nuffield Department of Medicine in Oxford, decided to try and find the gene for adult polycystic kidney disease (APKD). This is a dominantly inherited illness that results in kidney failure in middle age. Although it is quite rare in the general population, it is very common in people who require dialysis with an artificial kidney, or renal transplantation.

Reeders became famous in Oxford, when, as a junior hospital doctor, he would telephone his stockbroker during professorial ward rounds. Thus, from early in his career, he was notable as the possessor of exceptional organizational ability. Having decided to go hunting, he single-handedly collected blood for DNA from most of the known families with APKD in England. He set up his laboratory bench in the Department, extracted the DNA, wrote round for markers, and started looking for linkage.

In the laboratory next door to Reeders was a haematologist called Doug Higgs. Doug studies the genes for haemoglobin, the oxygen-carrying pigment in red blood cells. One haemoglobin gene is called alpha-globin, and it is found on chromosome 16. Next to the alpha-globin gene was a highly variable region of DNA called a minisatellite, which makes a very good probe for gene hunters (see page 39). When Reeders wandered into the alpha-globin lab and asked if they had any probes, it was only natural that Higgs should give him the alpha-globin minisatellite.

Reeders tested the probe on his families. It was linked to APKD in all of them. There was absolutely no doubt that he had located the APKD gene. Reeders denies that the alpha-globin minisatellite was the first probe that he tested, although the belief in the Department is that it was. Even if he were lucky, it was nevertheless an outstanding piece of work: within a year of starting his project he had found that APKD was on chromosome 16. Within a few further months grants were being offered to him on a plate, and a few months after that he had joined the 'brain drain' to the United States.

Reeders and his expanding group rapidly found flanking markers around APKD, and quickly zeroed in on a relatively small region that must contain the gene. There the luck ran out. The region was full of genes. At last count, after five years of considerable effort, there are twenty genes in the space of a few hundred thousand bases. Other scientists are competing for the prize, and the funding bodies are agitating for success. Finding a gene is one thing; finding mutations and proving it is *the* gene is another thing altogether. To sort twenty genes for a mutation that may only be a single base pair substitution can only be accomplished by mind-numbing, thankless, repetitive work; months and years of work through which the researcher hopes for the lucky break that may never come.

Greg Germino, an American who worked in Reeders' group, has an analogy for gene hunting. He says that to attempt to find a gene is to put yourself in the circumstances of the heroes of Conrad's *Heart of Darkness* or of the soldiers in the film *Apocalypse Now*. You start happily at the river mouth with your great adventure before you. As you move upstream, and the going gets harder, and the natives steal and cheat, you begin, at first without noticing, to become corrupt. The going gets worse and worse, and you become tired and ill-tempered, and the natives retaliate, and so do you, and the atrocities mount. At the very end you are stuck in Hell, mumbling 'The Horror, the Horror' and hoping against hope only that someone else will clone the gene and take you out.

There are thousands of other single gene disorders, caused by a whole bestiary of bizarre and wonderful genes. The discovery of each will add something quite new to our understanding of what we are and how we work, but they are rare. As a general rule in medical science, the effort is not justified when the number of researchers investigating an illness outnumbers the number of people with the disease. The heroic efforts needed to capture and clone a mutant gene, at least with the current state of technology, means that most will have

to wait. One other single gene was worth hunting, although not everyone would consider that it caused disease: the gene was that which
causes the odd condition of maleness.

CHAPTER 7

SEX

Sex is obviously genetic. We are, with very few exceptions, born either male or female. Environment or experience does little to change the physical facts of our birth. But before retracing the search for the genes that control our sex, there are some difficult questions that can be asked. What, for example, is the purpose of sex? Why do we come in two sexes? What evolutionary force has divided so many forms of life into male and female? The answers are not straightforward.

Sex is a major preoccupation of our species: men think about it every eight minutes; women, if we are to believe the magazines left by them in aeroplanes, think about nothing else. Sex is the subject of its own industry, and yet the font of much that is greatest in the arts. It is not surprising then that the word 'sex' conjures up a multiplicity of images. If you asked any number of people what they thought of sex, you would get the same number of different responses. However, if you asked them what, in their opinion, scientists thought of sex, the answers might be rather more uniform, and certainly more prosaic. The biological reasons for sex, most would maintain, were for reproduction. Without sex there would be no children, and without children there would be no human race. Indeed, without sex there would be very little life on earth.

Or would there? Birds do it, trees do it, we do it, but not every form of life does it. Bacteria, those indefatigable bundles of inventive DNA, don't need it. Many of the bacteria swarming on your skin will have divided in the time you have taken to read this page, and they will have had no recourse to sex. The humble amoeba can divide away, happily making identical copies of itself for ever. A bisected hydra will happily grow two new halves to replace those lost. Many plants can be propagated by cuttings, bypassing entirely the messy business of fertilization. Why, then, does sex exist? The reasons are quite complicated; sex is much more than reproduction. Sex is about evolution and about conflict.

A simple organism, such as a bacterium, keeps its genes on a single chromosome. Before the bacterium reproduces, the chromosome copies itself. A membrane grows between the original chromosome and the duplicate, dividing the bacteria in half. There are then two identical bacteria, each with a set of identical genes. This elementary replication can continue indefinitely, producing more and more identical copies of the first bacteria.

A difficulty only arises when chance inserts a mutation into a gene. Mutation is the fuel of evolution. Bacteria are not to be thought of as interesting little organisms, subsisting by and large in pools and ponds. Bacteria are everywhere; waging total warfare, with each other and with every other life form on earth, competing for every single micrometre of space and microgram of nutrient. Mutation, although usually deleterious, very occasionally gives a bacterium an opportunity to do better than its fellows.

Conflict breeds efficiency: for a bacterial species to be inefficient is for it to perish, crushed by the weight of ruthless competition. Evolutionary efficiency is improved by two strategies. The first is for an organism to keep more than one copy of each gene. In this way a deleterious mutation can be tolerated because the second copy of the gene may carry on as normal. The mutated gene is then freed to evolve into something new and useful.

A second strategy for efficiency is that of genetic mobility. Even in the simplest forms of life, genes are never alone. A mutant gene is always bound to the same companions on the same stretch of DNA. If the mutant is deleterious, then all the companion genes will suffer. Conversely, a useful mutant may be unable to succeed because it is trapped on the same DNA as a particular set of other genes. From the gene's point of view, this is the equivalent of being born into a bad neighbourhood.

For our understanding of the gene's point of view, we must be grateful to the 'Oxford School' of evolutionary geneticists, and in particular to Richard Dawkin, who put forward the concept of the 'selfish gene'.

Prospective undergraduates at Oxford University are required to pass an interview. To test the mental fettle of the candidates the interviewers can set puzzles. These often take the form of a mythical jungle, or a prison, in which there are two people or two types of creature. The

creatures are called cooperators or defectors. Cooperators are concilia-
tors and defectors are aggressors or cheats. Which type of creature, the
perspiring interviewee has to decide, is it better to be? The point of
such a dilemma is that there is not a right answer: in a world of cooperators it is an advantage to be an aggressor; in a world of aggressors
there might be advantages either way.

It is a short step from a hypothetical aggressor in a mythical forest
to a selfish gene in the jungle of life. The Oxbridge interview thus may
have led the evolution of the Oxford school of evolutionary genetics.

The thoughts explored in an Oxford interview belong to a branch of
science known as 'game theory'. Game theory was invented by Johnny
von Neumann, a mathematical genius who emigrated to the United
States from Hungary. Von Neumann had all the traditional trappings
of genius: he had a photographic memory, he could speak to his father
in classical Greek at the age of eight, he was a womanizer, a renowned
party-goer, and an appalling driver – one crash-prone intersection at
Princeton was christened 'von Neumann Corner'. Von Neumann pioneered the modern computer as well as game theory.

Game theory tries to find the best strategy for a game player in a
given set of circumstances. In America after the war, the game that
preoccupied the Pentagon was nuclear war. The Russians and the
Americans both had the capacity to destroy humanity. How should
the Americans react to Russian aggression without precipitating catastrophe? The American government contracted out the investigation
of this and other problems to 'think tanks' such as the RAND
Corporation, for whom von Neumann worked. This right-wing think
tank considered the unthinkable: how to conduct a successful nuclear
war. At one stage during the Cuban missile crisis the RAND
Corporation advised President Kennedy to bomb the Russians.
Happily he desisted. It is not a surprise that von Neumann may have
been the model for Stanley Kubrick's Dr Strangelove (although other
candidates, such as the Hungarian-born bomb-maker, Edward Teller,
exist). However dangerous it may have been in the real world of
nuclear politics, game theory, as propounded by von Neumann, was
to prove a fertile inspiration for researchers in fields as disparate as
artificial intelligence and genetics.

By applying game theory to genetics and, in a revolutionary
approach, considering the process from the point of view of the gene
instead of the whole organism, Dawkins could clarify facets of evolution
that previously had not been understood. He intended the term 'selfish
gene' to establish the gene as the centre of the evolutionary process.

After Dawkins popularized the epithet 'selfish' in relation to genes, it became evident to others that not only were genes self-seeking in general, but also there were genes which were genuinely and exceptionally selfish; that is they could escape the rules, whatever the rules might be, and replicate at the expense of other genes. From selfish genes it was a short step to 'ultra-selfish genes', a term invented by James Crow at the University of Wisconsin. In the pursuit of their own ends, ultra-selfish genes have the potential to destroy a species. The Oxford evolutionary geneticists have now carried the idea of a selfish gene to its logical conclusion: it is every gene for itself out there, dog eat dog, brother eat sister, a Thatcherite Utopia.

It should not come as a surprise to be told that sex is a process that involves many elements of selfishness. Lawrence Hurst, also from Oxford, has explored how sex may have evolved through the selective action of selfish genes. The story begins with bacteria. Although bacteria most often replicate without sex, they are not completely asexual; they can exchange genes to some extent. Or, from the Dawkins standpoint, the genes can change bacteria.

A bacterial chromosome divides in an all or nothing way. As a group, its genes can only be copied into a new bacterium with identical characteristics to its parent. Some bacterial genes however survive free. They exist independently from the main chromosome, in plasmids (see 'The Modern Map', page 40 ff.). Plasmids can divide without the constraints imposed on the main chromosome, and can exist in multiple copies within the bacterium. Plasmids can also move between different bacteria, particularly if there has been some damage or stress to the bacteria.

Plasmids first came to scientists' attention because they can carry the genes for antibiotic resistance, known as R (for resistance) factors. By swapping R factors, a population of bacteria exposed to antibiotics can build up antibiotic resistance very quickly. Starting from a single source, the same plasmid can eventually be found in quite unrelated types of bacteria. The enormous and growing problem of bacterial resistance to multiple antibiotics dramatically illustrates the evolutionary usefulness of plasmids – at least to the bacteria that carry them.

The genes in plasmids can also swap in and out of the main bacterial chromosome. In theory, a plasmid can pick up any gene it likes, and move it to any other bacterium or bacterial chromosome. This is an enormous advantage to the gene in the plasmid. A gene confined to a chromosome, even if it confers a spectacular benefit to the bacterium carrying the chromosome, is limited in its expansion to the direct descendants of that bacterium. If it can move sideways into many members of its own generation, then its numbers in subsequent generations will be enormously increased. Similarly, the Sultan with his many wives and many children ensures that his genes will be more widely propagated than those of his monogamous subjects and thus increases his chance of genetic immortality.

Even after the discovery of plasmids, it was assumed that this sort of promiscuous behaviour only occurred in bacteria. However, Barbara McClintock found genetic elements that could move freely about the maize genome (see 'DNA', page 15 ff.), elements which are now known as 'transposons'. Transposons are not genes in the usual sense because they do not make proteins. Elements in their peculiar structure, which are either borrowed from viruses, or which later evolved into viruses, allow the transposons to insert themselves into the DNA, or to excise themselves and move somewhere else. Transposons can alter gene function by disrupting the normal processes of control or transcription of a gene. Like plasmids, they may also pick up genes in their travels, dropping them off in unexpected places with unpredictable results. Although transposons may influence evolution, their main role is to copy themselves, often at the expense of the genes they disrupt: they are genuinely selfish.

Some plasmids (known as F plasmids) take sex a little further. In conditions of environmental stress the genes on the F plasmid induce the growth of a pilus, or tube, from the wall of the host bacterium. The tube enters the wall of a neighbouring bacterium. The F plasmid duplicates, and one copy moves down the pilus to enter the other bacterium. The F plasmid has then successfully ensured that it has twice as many descendants as the plasmids that do not induce pilus formation. This is another example of a selfish gene. The F factor is subverting the apparatus of the cell for its own ends, with no obvious advantage to the donor, or the recipient, bacteria; indeed the F factor causes the donor some disadvantage because it has to expend energy growing a pilus.

The F factor has an even more interesting effect when it has been assimilated into the main bacterial chromosome. When this is the case the pilus forms as before; however, the entire bacterial chromosome replicates itself, instead of just the plasmid, and one copy passes down the pilus to invade the neighbour. In the neighbour, many genes from the invading DNA have the chance to insert themselves into the host genome.

In this case, although the action of the F factor remains selfish in that it has succeeded in copying itself into another bacterium, the actions of the selfish gene have affected many other of the bacterial genes as well, with potential benefit for all of them. Perhaps in these circumstances we might refer to 'Attila genes', which launch invasions, or 'Moses genes', which lead their companions into the Promised Land.

Bacteria can shuttle genes around fairly easily because they are not very complicated. As more complex organisms have evolved, the process of exchanging genes has, necessarily, become very carefully controlled.

Above the bacteria on the evolutionary ladder are organisms in which the DNA is kept apart from the rest of the cell. Robert Brown was a Scottish botanist who, in 1831, first noticed through his microscope that cells had a large central structure. This region is the nucleus. The nucleus contains DNA wrapped in a membrane and, instead of just one chromosome, there can be many. The main substance of the cell, in which the nucleus is suspended, is called the cytoplasm. Organisms with this type of cell are 'eukaryotic', which means 'having a real nucleus'. Cells without a nucleus, such as bacteria, are 'prokaryotes'. Eukaryotic cells did not evolve in a direct way from prokaryotes. The nucleus was probably formed when one prokaryote invaded another. Genes in the invading organism took control of the cell, and eventually ejected or assimilated that cell's DNA.

Eukaryotes have an advantage over bacteria because they have two copies of each chromosome in their nucleus, and therefore two copies of each gene. The evolution that led to the duplication of each chromosome must have been far from simple. Perhaps the second set of chromosomes came about when one cell invaded another. This type of primitive sex can still be seen in some moulds, such as corn smut, a fungus which infects wheat. During the sexual phase of this organism the genes from each parent are present in each cell, but remain in separate nuclei. Regulation of cell division with two independent nuclei must be of horrendous complexity and the advantage of both sets of chromosomes sharing the same nucleus is clear.

A single set of chromosomes in a cell is called the haploid number, a double is called diploid. The extent of the coexistence of two sets of chromosomes in the life cycle of organisms is very variable, reflecting the evolution of diploid cells from haploid. Some organisms, such as simple plants and many parasites, spend much of their existence with a solitary haploid set. We, on the other hand, spend most of our existence with two: only the sperm and the ovum before fertilization have the haploid number of chromosomes.

More than two sets of chromosomes are possible: some higher plants can be extraordinarily relaxed about the number of chromosome copies they carry, having four or eight or even more sets. The flexible number of chromosomes means that plants can very easily form hybrids from parents of different species or varieties. This flexibility is of a huge evolutionary advantage to plants, and a great help to plant breeders.

Cells with the diploid number of chromosomes in two nuclei were successful because they had two copies of vital genes. They nevertheless suffered the same disadvantage as sexless bacteria: their genes were linked to the same partners on the same chromosome in perpetuity. However, once the two sets of chromosomes shared the same nuclear membrane then it became possible to exchange genes between chromosomes.

In bacteria with an F factor, after the invading chromosome has advanced down the pilus, the exchange of genes occurs with the two chromosomes together in a scrimmage. At the end only one chromosome remains. In eukaryotes the two sets of chromosomes, one from each parent, spend their existence as if they were still in separate nuclei: when the cell divides the paternal and maternal chromosomes divide separately, and each give a copy to each new cell. The only exception occurs when the sperm and the ovum are formed. These cells contain one copy only of each chromosome.

Because the result of this type of cell division ends with only half the number of chromosomes it is known as reduction division, or 'meiosis'. Before reduction division the chromosomes duplicate, as they do in normal division, but instead of going their separate ways they now line up side by side. An extraordinary thing then happens: the paired chromosomes break at certain points and form 'crossovers'. A completely new pair of chromosomes is formed, made of genes originally from the father mixed with genes that originally came from the mother. One pair of the new chromosomes is extruded, the other become ready for mating, in the nuclei of sperm or ova.

The mixing process of meiosis is a critical step in allowing genes their freedom. Without it, the next generation would receive either a grandfather's chromosome or a grandmother's, and their genes would be exclusively grannie's or grandpa's. After meiosis, however, the new chromosome can have a bit of both, and the chance of finding an advantageous mixture is much higher. The effects of meiosis explain Mendel's law of random mixing of genetic traits. The positions of the crossovers, or recombinations, are the currency of the gene mapper.

Mapping simply resolves the position of a gene to one side of a recombination event (see earlier chapters).

Meiosis has a weakness. James Crow has shown that it allows selfish genes to 'subvert and cheat' the reproductive machinery. Ultraselfish genes exist, for example on the fruit fly sex chromosomes. They suppress other sex chromosomes, so that they are preferentially transmitted to the sperm or ova. This causes all the progeny of a fly to be of the same sex. In the next generation the ultraselfish genes then have twice as many copies as their rivals on the other chromosome. Because they are so successful, these genes can spread rapidly through the succeeding generations of a population. However, if they are totally successful the flies will all eventually be of the same sex, leading to the immediate extinction of the species. Lethal as these genes are, they exist in many wild populations of fly. They probably carry out some useful function, the nature of which is quite unknown. As they have not spread throughout the fly population, Crow suggests that there are evolutionary mechanisms for removing the cheaters, to 'preserve the honesty of the Mendelian gene shuffle'.

So far we have seen how sex solves the requirements for evolutionary efficiency. It gives sexual organisms two copies of their genes, and it allows genes to move away from their companions on the same stretch of DNA or chromosome. Lawrence Hurst has also explored the next question of sex. Why is the number of sexes greater than one? If it needs to be more than one, why should there be only two sexes? Besides male and female, why not threemale and fourmale? Hurst explains these phenomena, inevitably, through conflict.

The nucleus is not the only invader of ancient eukaryotic cells. Eukaryotes have other spherical structures in their cytoplasm. These are mitochondria and chloroplasts. Mitochondria are small and full of densely folded membranes. They are the power stations of the cell, converting metabolic energy into a form that can fuel its functions. Chloroplasts are only found in plants, and capture the energy of the sun by photosynthesis. They are of a similar size to mitochondria.

Mitochondria and chloroplasts, known collectively as organelles, betray their parasitic origins because they retain their own DNA: each contains its own circular chromosome and its own genes. Just like plasmids in bacteria, the organelles and their DNA replicate independently of the nuclear DNA. This means that there may be two or three separate sets of genes in a cell. Each has its own origin and, according to Hurst, each has its own agenda.

Mating in eukaryotes is controlled by the nuclear genes. Before mating, the number of chromosome copies in the sperm and ova is reduced to one, the haploid number. This ensures the fertilized ova will have the normal number of two copies of each chromosome, one copy from the father, the other from the mother.

Competition between maternal and paternal genes in the new cell is resolved because the nuclear genes from each parent are equally represented. Thus neither parent's genes are at a disadvantage. This is not always the case, as we shall see in the next chapter, but it may be taken to be generally true.

The genes in the organelles are not, however, controlled by the nucleus, and are not bound by its rules for equitable sharing of the new cell. Hurst suggests that if fusion of two cells takes place in mating, then the genes in the organelles from the two parents are potentially in conflict – 'a tragedy of the common cytoplasm' as he describes it. Fusion forces cytoplasmic genes from one cell into the company of cytoplasmic genes from another. Because all genes are potentially selfish, they may break into warfare for the sole occupation of the cytoplasm.

This warfare is not in the interest of the nuclear genes, which initiated the mating. To prevent the war the nuclear genes must choose between one set of cytoplasmic genes and the other. Elaborate systems for favouring the cytoplasmic genes of only one parent have therefore evolved. The need to choose between cytoplasms means that each parent must have a separate identity. In other words, one parent must be male and the other female.

One means of achieving this is for the cytoplasmic genes of one parent to be left behind at the time of mating. Hurst calls this conjugatory sex. Most higher forms of life practice conjugatory sex. The sperm carries virtually no organelles into the ovum at conception. The mitochondria in all the cells of your body are derived solely from your mother; if you are male, you will not pass your mitochondrial genes to any of your children.

A second type of sex is that which Hurst names 'fusion sex'. In fusion sex, which occurs in simple unicellular organisms such as algae, the cytoplasm from both parents is mixed. To control potential conflict between cytoplasmic genes the nucleus from one parent can suppress its own cytoplasmic genes. This is done at an energy cost, so matings between two 'suppressors' are inefficient and, in evolutionary terms, will be disadvantaged. Matings between two non-suppressors are also inefficient, because of the possibility of warfare. Matings between a suppressor and a non-suppressor are of maximum efficiency. To ensure that suppressors always meet non-suppressors, a second gene 'chooser' is necessary.

With choosers and non-choosers, and suppressors and non-suppressors, we are again in the realm of game theory and the wood-panelled room with the Oxford interviewers. Surprisingly, in the real world these genes exist and are found in many varieties of life.

Cytoplasmic conflict has led to maleness and femaleness, but has not explained why there should be only two sexes. There are rare species which do have more than two sexes, such as the thirteen-sexed slime mould. This peculiar mould organizes its sex life in strict hierarchy: sex 13 cytoplasm dominates that of the other 12 sexes. Sex 12 gives way to 13, but rules the other 11 sexes below. Sex 1 is at the bottom of the pile, its cytoplasmic genes doomed to extinction.

The problem with this complex arrangement is that it is inefficient: it is too easy for the organization to break down, and too many nuclear genes have to work at suppressing the thirteen variants of cytoplasm. A mutation, for example, in sex 2 would allow its mitochondria to compete on equal terms with the 10 orders above it. The whole class system would collapse in revolution. Whenever there is a collapse, the system reverts to its most efficient form, that of two sexes.

Thus in sex, as in California, anything is possible. Sex is beneficial because it allows genes to move around and to combine with new genes. If a gene can find a way of taking advantage it will. Male- and female-ness may be the result of cytoplasmic war.

CHAPTER 8

LA DIFFERENCE

If the reasons for the existence of sex are not simple, then, given the obvious differences between men and women, and the difficulties they seem to have in understanding each other, it would be easy to imagine that the genes of men and women differ in hundreds or even thousands of ways. In fact, just one gene may be responsible for the entire mystery.

By 1910 it had been realized that female creatures had different chromosomes to males. Females seemed consistently to have a complete pair of one type of chromosome, whereas males seem to have just one of these chromosomes with no twin. This particular pair of female chromosomes was labelled the X chromosomes. Later a single, much smaller Y chromosome was detected in males. As we have already seen, the association of sex with these chromosomes was the first generally agreed linkage of a characteristic with a chromosome, and opened the door to the chromosomal theory of inheritance.

In T.H. Morgan's beloved fruit flies, sex seemed to be decided by the number of the X chromosomes. Two, four and six X chromosomes were found in females; one and three were found in males. But in the case of humans and mammals the link between combinations of X and Y and sex was not at all clear. Part of the confusion was because it was very difficult to get good preparations of human chromosomes to look at under the microscope. Chromosomes are only visible for a short time, in the moments just before a cell divides. For the rest of the life of a cell the chromosomes unravel into a loose mess called chromatin. The cells which divide most often in the human, and which might therefore be expected to show the chromosomes to best advantage, are the cells that make red and white blood corpuscles. These cells are in the bone marrow, which in 1910 was far from accessible. The early workers had to try and make sense of smudges of tissue that were only dimly visible at the bottom of a microscope. Nevertheless, the general size and shape of the human chromosomes were known by the end of the 1940s. Determining their number took a little longer.

In the 1940s it was believed that humans had 48 chromosomes. This number was included in all the standard textbooks, and any cytogeneticist (chromosome counter) worth his salt could identify the full 24 pairs in his preparations. Unfortunately, there are not actually 48 chromosomes in normal humans. George Klein tells, in his story *The Emperor's New Clothes*, how two researchers in Sweden, called Tijo and Levan, found the truth. By 1956 they had raised the art of chromosome preparation to a new plane. For the first time it was possible to make reliable counts of the number of chromosomes in many cells from the same individual. To their surprise, Tijo and Levan found that they always counted 46. It had not occurred to them before they started their experiments that 48 was wrong, but eventually they were forced to conclude that 46 must be the right number. They even found, in a very widely used textbook of cytology, a microscope picture from an acknowledged master of chromosome identification, T.C. Hsu. The figure legend claimed 'a typical complement of 48 chromosomes'. Tijo and Levan counted the chromosomes in the picture. There were 46! George Klein's title makes the moral of this story clear.

By 1959, chromosome counting was a commonly available skill. Nowadays it is a trivial, albeit painful, procedure to withdraw a sample of bone marrow. The cells are treated with a drug, originally used to treat gout, which stops cells dividing at the stage when their chromosomes are best formed and easiest to see. A number of stains can then be used to make individual chromosomes more easily identifiable by the pattern of bands on their surface.

The study of chromosomes led to the first steps towards discovering the difference between men and women. Appropriately, women have contributed as much to this field as men, although when the search began in the 1950s a successful woman scientist was a rarity.

Some syndromes of abnormal sex differentiation had been known for centuries. One of these was hermaphroditism, in which the sexual characteristics of male and female are combined in the same individual. Another was Kleinfelter's syndrome. People with this disease are usually effeminate boys who are very tall, don't shave, and have small testicles. People with Turner's syndrome, on the other hand, look female, but are short and infertile. For many years Turner's syndrome was thought to be a state of failed maleness.

Although females, and all their cells, have two copies of their X chromosome, both chromosomes do not need to work simultaneously. It so happens that in each cell one X chromosome is tucked up in a little bundle in the middle of the nucleus. This bundle can be seen under the microscope, and is called the Barr body, though, as there were two authors on the 1949 paper announcing its presence, it should be called the Barr-Bertram body. If you try and say 'Barr-Bertram body' repeatedly, you will realize why poor Bertram has lost his chance for eponymous fame. Barr and Bertram were not the first to observe the body, but they were the first to recognize that it was absent in normal males. It was, however, some time before the Barr body was recognized as a quiescent X chromosome: its absence was simply taken to be an indicator of maleness. Belief in this association was so strong that examination for a Barr body was the standard test for chromosomal maleness for many years. It was introduced by the International Olympic Committee in 1967 to catch out men competing as women in the Olympic Games, and remained in use until 1991.

When doctors looked at the cells of individuals with Turner's syndrome they did not see the Barr body. It was therefore assumed that Turner's syndrome was a state of incomplete maleness, and patients with the syndrome were called 'cases of sex reversal', or 'chromosomal males', or 'genetic males'. This, I am sure, was the source of considerable grief to unfortunate females who were so labelled. In 1959 a British paper by C.E. Ford and K.W. Jones showed that people with Turner's syndrome had 45 chromosomes instead of the usual 46. The only sex chromosome present was a single X. In geneticists' terms this is signified by XO. As patients with Turner's syndrome are anatomically female, this meant that the gene or genes causing maleness must be on the absent Y chromosome, and that femaleness was not caused by a double dose of X. I like this paper because of the touch of humanity at the end. The authors conclude,

...it should be emphasized that the XO patient should not be referred to as an instance of 'sex-reversal'... she is a female, with an abnormal genotype.

In the same year Patricia Jacobs and J.A. Strong in Edinburgh counted the chromosomes in a case of Kleinfelter's syndrome. The patient was a man, in whose cells they found a Barr body, meaning that he should be female, but there were 47 chromosomes, rather than 46. As well as two normal Xs he had a normal Y. His two Xs had not caused him to become female, which was further evidence that maleness was conferred by the Y chromosome.

The scientists who examined these two cases had drawn very important conclusions from apparently simple observations. Turner's and Kleinfelter's syndromes are far from rare and were well described in the medical literature. However, no-one had previously thought that they might provide concrete information about the development of maleness. It is in this capacity for finding the novel amongst the apparently commonplace that the scientist and the artist exhibit the same trait: creativity.

Patricia Jacobs continued to look at sex chromosomes. Over five years she and her colleagues examined the chromosomes of several thousand people. From all these subjects she found only four abnormal Y chromosomes. Two of these were from males, and, surprisingly, two were from females. The 'X-Y females' were missing two things: male sexual characteristics, and the short arm of their Y chromosome. The males, on the other hand, maintained the short arm, although other bits of the chromosome were missing. Jacobs deduced that the male-determining genes must have been on the short arm of the Y chromosome, and that the genes on the long arm were not primarily concerned with sexual differentiation.

The science of sex-determination then shifted its attention to fishes and amphibia. These cold-blooded creatures are interesting sexually because some of them can change their sex according to what is happening in the environment. Jones and Singh, also from Edinburgh, pointed out that this meant that the capacity to develop ovaries or testes is shared by both males and females; what differentiates the sexes appeared to be a switch which set a pathway either to male or to female gender development.

Jones and Singh studied the chromosomes of snakes. Some snakes, such as boa constrictors, do not have sex chromosomes, although they do have sex-determining genes on other chromosomes. Other snakes, more evolutionarily advanced than boas, do have chromosomes which vary between sexes. These are called W and Z, because they are not quite the same thing as X and Y. Jones and Singh attempted to isolate the DNA sequences from the snake W chromosome, to see if it corresponded to mammalian sex-determining DNA. This, for a variety of reasons, was unsuccessful, but the idea that a single switch might determine the difference between males and females was by then firmly fixed.

The search then turned to a protein which, in mice, was found only on the cells of males, and which seemed to be derived from a gene on the Y chromosome. H-Y is of a type of protein called a 'histocompatibility antigen'. These proteins form a complex with sugar molecules, and are found on the surface of cells. They are part of the very complicated mechanism by which the immune system recognizes that which is self and that which is not. During the early growth of the immune system, proteins present on the cell surface are recognized to be self, and are treated as such thereafter. When non-self proteins appear, as occurs during virus infections, or after a kidney transplant, the cells are attacked by the immune system and destroyed.

Immune recognition does not seem immediately to have anything to do with sex, but the recognition of differences between two types of cells may have had an important function in primitive sex. It is possible to construct a theory of male- and female-ness on these lines for the H-Y antigen, and it was popular for ten years.

Sadly, Anne McLaren at the MRC Mammalian Development Unit in London then found male mice with testicles and no H-Y antigen. Like many of the best ideas in science, the notion that the H-Y antigen was the sex-determining factor thus died a lonely death. Sad though the death of hypotheses is, we should not mourn them too much. They are as perennial as the grass; they come up like the flowers, and are cut down. Even bad hypotheses can generate interest, and serve as facilitators for the appearance of new and better theories.

By 1986 the sought-for sex-determining gene in mice had been labelled *Tdy*, or testis-determining factor–y. In humans it was known as *TDF* – because, to most geneticists, humans are more important than mice, and not because they have bigger testicles.

By this point new techniques allowed much smaller pieces of chromosomes to be recognized and traced if they had 'translocated' to another chromosome. The hunt therefore returned to chromosomal abnormalities. Men with two X chromosomes were found and it was shown that they each had a tiny part of their father's Y chromosome hidden on another chromosome. As they were male, and as they had no other Y chromosomal material, it could be assumed that the difference between man and woman lay in the translocated Y chromosomal DNA.

Besides men with XX, a woman with XY was also found, who, it was assumed, must be female because she had lost the *TDF* from her Y chromosome. Taken together, the evidence from these abnormal chromosomes reduced the region that must contain *TDF* to only 140,000 letters of the genetic code. This is not a trivial amount of DNA

– 140,000 letters are about equal to a hundred pages of a book – but on average would be expected to contain only one gene. In 1990 a gene was found within this segment of DNA.

This gene was called *ZFY*: zinc finger y. Zinc fingers do not belong to a character from the Wizard of Oz; they are proteins with finger-like protrusions containing zinc. Zinc fingers can sit on DNA and control whether it is transcribed into RNA to make a protein. *ZFY* was therefore a very promising candidate for the TDF gene. Sadly, three men with XX and no *ZFY* were then described by Peter Goodfellow at the ICRF laboratories in London. At the same time a team led by Robin Lovell-Badge at the MRC National Institute for Medical Research in Mill Hill showed that *Zfy* in mice was not expressed at the right time or in the right place to be responsible for the development of testes. The candidate, as Anne McLaren put it, was not elected.

The chromosomes of the men without *ZFY* reduced the region of the DNA that should contain the sex-determining gene to 65,000 letters and further evidence from abnormal chromosomes then reduced the region to 35,000 letters. In gene hunter's terms, this is a very small region, so small as to make one doubt that there was a gene in there at all. The doubt was increased by the presence in the region of a large amount of repetitive nonsense sequences of DNA.

Many scientists claim the paradigm for the relationship between scientists and clinicians is that of Sherlock Holmes and his friend Watson. Holmes could well have been speaking of *TDF* when he observed to his assistant, 'How often have I said to you that when you have eliminated the impossible, whatever remains, *however improbable*, must be the truth?'

Sure enough, Goodfellow and Lovell-Badge found a highly improbable gene within the remaining DNA. The gene itself was small, only 250 base-pairs long, compared with, for example, the 200,000 bases of the Huntington's disease gene. The gene coded for a protein of only eighty amino acids, and was christened *SRY*, for 'sex-determining region of Y'. The sequence of the human gene was very similar to the mouse equivalent, sry, suggesting that its function was the same in the two species.

Final proof that this was the sought-for gene was carried out in the most satisfactory way possible. The *Sry* gene was inserted onto a mouse X chromosome. If the resulting transgenic mouse was male with an XX genotype, then *Sry* was the sex-determining gene. Goodfellow, renowned for his puckish sense of humour as well as for his science, and Lovell-Badge submitted the results of their transgenic experiment to *Nature*. The article was summarized by a picture of two mice holding on to a bar. The left-hand mouse was the possessor of an X and a Y chromosome, the right owned two Xs and the transgenic *Sry* gene. Both mice sported a visible pair of testicles. *Sry* was indeed the sex-determining gene.

However, as Goodfellow, now the Professor of Genetics at Cambridge, tells the tale in his public lectures, the reviewer to whom *Nature* sent the article objected. What proof was there, the reviewer asked, that the XX mouse with the testicles was indeed a male? As the standard definition of maleness had, up to then, involved the presence of testicles, this was unhelpful in the extreme; but Goodfellow and Lovell-Badge were able to prove the masculinity of the transgenic mouse in an unexpected way: they put it into the same cage as a female mouse in oestrus. The transgenic mouse squeaked, as male mice do before mating, and enthusiastically mounted the female. When introduced to other females in oestrus the mouse behaved in the same way. Goodfellow and Lovell-Badge wrote back to *Nature* describing the results of the experiment. They pointed out that not only did the transgenic mouse consider itself male, but that the female was also obviously of the same opinion. The paper was then accepted.

Extraordinarily, *SRY* was found to resemble a gene in yeast which makes a mating protein, called Mc. From yeast to man is a long way to travel, implying that the early mechanisms for differentiating organisms into male and female were so effective that evolution has not replaced them.

La difference, all that separates men from women, is therefore only 250 base pairs of DNA which produce a protein containing eighty amino acids. All the other, very obvious, symptoms of maleness stem from the presence of testicles, and the subsequent all-pervasive action of the hormone they produce, testosterone. Seen in these terms, arguments about the cranial capacity or intellectual abilities of men and women seem a little nonsensical, as the genes for these attributes are identical in both sexes. It also seems odd to deny women the right to be priests, simply because they do not express an 80-amino-acid protein for a few hours in the first days of life. Women should also now be more sympathetic to men: we are the same beings except that to be male is to be awash in testosterone, aggressive and irrational to the last.

Although vigorous debate about real or supposed differences in ability between men and women is likely to continue indefinitely, there is little doubt about one thing: men are better at sport. The key to men's sporting advantage is obviously that they are bigger and stronger than women. Men's big muscles are due to testosterone. Before puberty boys and girls are not greatly different in their sporting ability, and restrictions on the sports that girls play are largely culturally imposed. All over England, for example, tomboys play soccer and rugby happily enough until nine or ten. After this, testosterone production begins in boys, and they become recognizably stronger than the girls. Two years later, when the testosterone storm of puberty begins, the boys become much stronger, and much more aggressive. By this stage girls can no longer compete physically with males of their own age.

The effects of testosterone, or its artificial equivalents, the anabolic steroids, are known to all sportsmen and women. Those who take steroids gain an advantage over their competitors. This is unfair, and so sporting associations test for steroid use and sometimes even ban those who have been caught out by the tests.

Women who use steroids do so to become more like men. If steroids were given to females in low doses before puberty and in high doses thereafter, the gap between male and female athletic performance would probably be very small indeed. Even with 'conventional' dosage regimes it is plain how male in appearance some female athletes become. It seems to me that the steroid-users are obvious in the line-up of any female international sprint competition. They have very little subcutaneous fat and their muscle definition is extraordinary. Before the age of steroids, women like this did not exist. Even the biggest muscles were clothed in some feminine fat, as is shown by the women in the same sprints who do not use steroids. Men who take steroids are less obvious, because they just look more like men. The exception is that their testicles are often tiny, a result of the body shutting down its own testosterone production in a vain attempt to restore normality.

Before steroids existed in a form that could be injected, some resourceful individuals realized that one way to gain the advantage of maleness in female competitions was actually to be a man. A need for sex testing to prevent men competing in women's events became clear as a result of several widely publicized instances when the winners of female competitions were subsequently found to be men. The winner of the 1932 women's 100 metres Olympic sprint final was discovered, after a fatal shooting accident in 1980, to have had testicles. The world high jump champion in 1938 was found to have both male and female sex organs, and was excluded from subsequent competition. Two women, half of the relay team that finished second in the 1946 European Championships, subsequently had sex changes. One of them even became a father. The winner of the 1966 women's World Downhill Ski Championship also rounded off her career with surgery and fatherhood.

Even before these examples became public, it was accepted that men had to be stopped from competing as women. The British Women's Amateur Athletic Association made it known in 1948 that women competitors, before they were permitted to take part in sport, should have a doctor's note attesting to their femininity. More pragmatic nations were less inclined to take these sorts of thing on trust, and so in Budapest in 1966, before the women's championships, women competitors were required to be inspected by a panel of three female doctors. In Kingston, Jamaica, in 1966, before the Commonwealth Games, a gynaecologist performed a manual inspection of the athletes, and the external genitalia of all female athletes were inspected before the Pan American Games in 1967.

If you imagine the athletes were angered or frightened by these events, then you are right. One highly unfortunate woman had two populations of cells in her body: these contained either XX or XXY. She won the gold medal for the 100 metres sprint in Tokyo in 1964, and held the world record over that distance. In 1967 she failed an inspection of her external genitalia, and it was found she had had an operation to remove undescended testes. She was publicly banned from athletics thereafter, and her name was expunged from the record books. One can only imagine her grief, as it is highly unlikely that she had ever considered herself as anything other than a woman. The uproar over this and other cases made the International Olympic Committee consider other means of sex testing female competitors. The IOC decided on an examination for the Barr body, as the understanding at the time was that it always indicated femaleness. This assumption was in error, as individuals with XXY, such as those with Kleinfelter's syndrome, will have both testes and a Barr body. However, had the woman with both XX and XXY cells been tested under this regime, she would have been found Barr body positive and allowed to compete.

Because of this problem, the IOC abandoned Barr body screening in 1992 in favour of an assay for the testis-determining gene which, as we now know, is the exclusive determinant of maleness. But even this test is not without its problems, as a condition exists known as XY femaleness. XY females make up the majority of athletes who are now excluded from competing as women – about one in five hundred among elite female athletes. These individuals have male chromosomes, but either have mutations in the SRY gene or lack the cell receptor which responds to testosterone. They grow up as normal females, except that they may be somewhat taller than average.

The Professor of Pathology at Cambridge, Malcolm Ferguson-Smith, has done much to clarify the difficulties involved in establishing fair sex tests. He points out that XY females gain no advantage from their Y chromosome, as their muscles are incapable of responding in a masculine way to testosterone. Even if they took anabolic steroids they would not respond to them, and Ferguson-Smith argues that XY females should not be excluded from female competition. He makes the sophisticated point that sporting excellence is conferred by genetic variation, and that it is unjust to exclude some individuals from sport because they possess particular genes, while including others because they possess other genes. People with 'abnormal' sexual states only differ from 'normal' by the smallest fraction of their DNA. Given the extent of natural variation it is perhaps sounder to consider these individuals as representing the extremes of genetic diversity.

There is a further inconsistency in sex testing which has not yet been addressed. As a woman has apparently never been successful in competing as a man, it has been considered unnecessary to call upon men to prove their gender, or their chromosomal status. This may not be a valid assumption. An extra X chromosome, for example, might give an advantage in height, and an extra Y, which is found in excess in male prison populations, might also confer an advantage in certain circumstances.

Sport apart, carrying extra chromosomes is not always an advantage. The sex chromosomes pose a particular problem. The X chromosome has many genes and the Y chromosome only a few. Those of you blessed with two Xs will have one more copy of the X genes than you need. Genes can be like medicine prescribed by a doctor. Twice as much is not at all the same thing as twice as good.

Two or more functioning copies of genes are sometimes necessary; an example is the genes for some chains of haemoglobin, the red pigment in blood. This is because the red cells need to be stuffed as full as possible with haemoglobin. However, stuffing other cells with a double dose of other proteins is likely to do more harm than good. For example, the genes which cause cancer can make entirely normal proteins in highly abnormal abundance. In Down's syndrome there is an extra chromosome 21, so that the difficulties of the afflicted children are due to an excess, rather than a loss, of chromosomal material. Many proteins are made up of separate chains from different genes which can be on different chromosomes: if production of the chains is unbalanced, the chain in excess may clog up the cell's machinery. To prevent this happening with X chromosome genes, a mechanism has to exist to suppress the extra X.

Because the Barr body was present when there were two X chromosomes, and absent when there was only one, it was inferred that the Barr body was the second X chromosome, condensed and isolated from the rest of the DNA. It was thus reasonable to assume that the genes in the condensed chromosome were inactive: only one X was working at a particular time in any one cell. But which X was turned off? The intriguing possibility existed that the inactive chromosome might always come from the father, or always from the mother: that is to say, the male and female inheritance to the child might not be the same. This would be a genetic equivalent to the sins of the father visited on the daughter.

A mouse geneticist, Mary Lyons, solved the problem of 'which X' by inspired observation. In mice, various mutations of coat colour were X-linked; in other words, the defective genes causing the variants of coat colour were to be found on the X chromosome, just as Morgan had shown that eye colour was traceable to the X chromosome of his fruit flies. Mary Lyons recognized that female mice with an X-linked mutation that affected coat colour were mottled, or dappled. They were not one colour, or the other, nor an even mixture of the two. The mottling was random, and not in a regular pattern.

From the pattern of dappling, Lyons deduced that one of the two X chromosomes was inactivated early in development, when the number of cells in the embryo was less than a hundred. The billions of skin cells were descended from ten or twenty of the hundred in the embryo when an X in each cell was switched off. In these twenty cells it would be entirely random which X was switched on and which was switched off: some would have an active X containing the mutant, and others an active X containing the normal gene. After embryonic growth was complete, the descendants of a particular cell would be together in the same patch of skin. A patch derived from a cell in which the mutant X was active would be the mutant colour, and the descendants of cells with normal X would be a normal colour.

The random pattern of the mouse's dappled coat meant that X inactivation was at random. This in turn meant that any woman is a mosaic of cells with either one or other of her X chromosomes active. The idea of two populations of cells in the same person is a little strange: it is almost as if there are two different people sharing the same body. This process is now known as Lyonization. The same thing holds for most of the genes on the X chromosome, and not just those affecting coat colour, and it holds for mammals other than mice.

Mary Lyons concluded her original 1961 paper with an observation on cats. Tortoiseshell cats have a distinctive black and yellow coat. Cat lovers know that there is no such thing as a male tortoiseshell: they are always female. Lyons suggested that the tortoiseshell cat carries two genes for coat colour, one black, and one yellow; one of these two genes is on each of the X chromosomes. In patches of skin where the yellow gene is on the active chromosome, the coat colour is yellow. Where the black gene is active the coat colour is black. It is the random activation of 'yellow' and 'black' X chromosomes which gives the cat its tortoishell coat. Normal women are likewise a mosaic from the different effects of their two X chromosomes – the only difference is that the pattern cannot be seen.

It is thirty years since Mary Lyons explained how the cat got its coat, but it is still not understood how a whole chromosome can be turned off. A gene called Xic (for X-inactivation centre) has been found in the right place for the inactivation focus, and is expressed only from the inactive X. How it works, no-one yet knows. The answer, when it comes, is likely to increase greatly our understanding of how genes are switched on and off, and to offer the potential for new types of treatment for some genetic diseases.

X chromosome inactivation and mosaic females are curious phenomena, but even more curious things happen with the male and female side of our genes. Until recently, geneticists believed that, except for genes on the sex chromosomes, the genes we inherit from each parent carry an equal weight. Science, however, teaches us that assumptions are never valid until they have been tested experimentally, and sometimes not even then.

The assumption that our genetic inheritance from our mothers is equal to that from our fathers was at first borne out by experiments in frogs. The experiment involved taking the nucleus out of a fertilized egg from a female frog. This left the cell cytoplasm primed with all the materials necessary to begin the development of a frog embryo. The only elements missing were the genes from the DNA in the nucleus. Without instructions from genes, the cytoplasm could do nothing, and the cell would quickly die. A nucleus was then taken from a skin cell of an adult frog, and placed into the empty cytoplasm of the egg. As all nuclei in all cells of the body contain copies of all genes, it was theoretically possible that a completely normal frog would develop from the reconstructed egg.

In fact the experiment worked brilliantly. An embryo grew into a tadpole, and the tadpole into a normal frog. This frog was an absolutely identical copy of the frog from whose skin the nucleus had been taken. A hundred frogs could be made in this way, each an exact replica of all the others and of the frog who donated the nucleus.

These carbon-copy animals are what most people consider to be clones, although, as we have seen, 'clone' has a much narrower meaning to geneticists. The implication of this tampering with frogs was obvious. If it was possible to make hundreds of clones from a single smug frog, then surely it was possible to make hundreds of copies of mice, and even hundreds of copies of human beings. No military dictator or female novelist need worry about their immortality; they could be replicated as perfectly as a film or CD is copied from the original.

Very sadly, but perhaps rather fortunately, when the experiment was tried in mice it failed. The embryos would develop some way towards maturity, but would eventually die because of a multiplicity of abnormalities. This was first explained by the need to have the nuclear genes in a particular state in the fertilized ovum.

In mammals an enormous amount of preparation has to be carried out by very specialized genes before and immediately after conception. These changes give the egg a front and back, a top and bottom, and a left and right. A pattern of genetic signals decrees that the head of the embryo will be there, and only there, and that the backbone will be here, and only here, even before the sperm has entered the egg. Although the pattern may be relatively simple at first, it is the fundament for the miracle of organization that follows. The signals can only be set up correctly if the ovum has developed in a specific way. The genes for the next critical steps in embryonic growth are also primed and ready to respond to conception. A nucleus taken from the skin of a mouse does not have its genes in the appropriate state of awareness, and the cloning experiment fails. In amphibia, such as frogs, the mechanisms for embryonic growth are much simpler, and cloning works.

Thus, although it was impossible to clone mice in the same way as frogs, there seemed to be good reasons why it did not work. However, there is more to the story, because the abnormalities in the mouse embryos formed recognizable patterns.

Geneticists are very interested in developmental abnormalities because they point to the action of particular genes that must be necessary for normal growth. It was possible that the abnormalities in the mouse embryos after nuclear transplantation might help trace genes that were critical to embryonic development. Very oddly, it was noticed that when the nuclei came from female tissues the membranes and placenta of the developing embryo were underdeveloped, but that the embryo itself was relatively normal. When the nuclei came from the father's side, then the foetus was small but the placenta was large and normal in appearance.

A very similar phenomenon is seen in plants. Plant geneticists had known early in the century that the male and female sides of a plant may have distinct genetic properties. In flowering plants a part of the seed (called the angiosperm) performs a similar function to the human placenta: it extracts nutrients from the mother plant to enable the embryo in the seed to grow. The genes from the male side seemed to control the growth of the angiosperm, and the genes from the female side of the plant controlled the growth of the embryo.

The origin of this father–mother effect was traced further by looking at mice which had inherited both copies of a particular chromosome from one parent. This can occur when an extra copy of a chromosome accidentally hitches a ride in the sperm or ovum. Sometimes, but not always, this results in abnormalities. The type of abnormality depends on the chromosome involved. What was surprising was that the abnormalities differed according to which parent had given the duplicated chromosomes.

An idea began to grow that genes which originated from the mother and those which were from the father differed in the way they were turned on or off in the growing embryo. It was as though male- and female-ness were in some way marked onto the chromosome, or the genes it carried, and that the mark stayed on the chromosome thereafter.

But mice were one thing and humans another. Moreover, no-one really understood what these odd findings meant, and the whole business of both chromosome copies coming from a single parent was artificial in the extreme, and the facts were not thought to be important by human geneticists. However, a paediatrician called Angelman, now in retirement on the south coast of England, had in his prime described a syndrome that he christened 'happy puppet syndrome'. The unfortunate children who developed the syndrome were mentally retarded, and had a characteristically odd and inappropriate laughter. In these more enlightened times we realize that it would be unnecessarily upsetting for the parents of a child to be told that it suffered from happy puppet disease, and the syndrome now bears Dr Angelman's name.

Angelman's syndrome is mercifully rare, and it would not be considered of great interest except for the fact that it is associated with another congenital illness, known as Prader-Willi syndrome. Children with the Prader-Willi syndrome also suffer from abnormal mental development, but they are plump and sleepy, and not at all like children with Angelman's syndrome. Prader-Willi syndrome is also quite rare, and would also have languished in the small print of medical textbooks except that it appeared in the same families as Angelman's syndrome.

Chromosomal abnormalities, involving the same part of chromosome 15, could be seen in families that carried both these diseases. It was years before it was realized that when the abnormal chromosome was inherited from the father, the child would have Prader-Willi syndrome, while when the mother was the source of the chromosome the child developed Angelman's syndrome. The two diseases were occasionally seen in the same family: Angelman's syndrome and Prader-Willi disease alternating in the same pedigree according to which parent had passed on the defective chromosome.

It was not possible to understand what this meant until systematic experiments on maternal and paternal chromosomes were carried out in mice. These found seven main chromosomal regions in which the parent of origin influenced the subsequent pattern of abnormalities in the mouse embryos. It was a mystery why the genes should behave in this way. The term 'imprinting', meaning 'marked', was coined to describe it because genes seemed to be marked to behave in different ways according to whether they had passed through the sperm or the ova on their way to the embryo. More specifically, the genes which are imprinted are switched off, perhaps in a similar way to X chromosome inactivation in females, while their counterparts in the corresponding chromosome from the other parent remain active.

Imprinting explains other odd things. Cancers and leukaemias (cancers of the white blood cells) are due to genes that have gone awry. Under the microscope, the chromosomes in cancer cells are often abnormal or broken and jumbled together. In some cancers the abnormal chromosomes have always been inherited from the mother. This finding is important because it may offer clues to new methods of treating these cancers.

The oddest thing of all, however, was revealed by the discovery of two genes in mice which are imprinted. The first gene (called insulin-like growth factor II) makes a protein that controls the growth of the placenta and thus the feeding of the developing foetus. The gene was shown to be maternally imprinted: the copy arising from the mother is inactive. Placental size is therefore controlled by the father's gene, and the mother is not permitted a say in how much nutrient she has to give to the foetus.

Growth factors are like hormones. Each growth factor or hormone has its own receptor, which is usually a protein or complex of proteins that sits on the surface of a cell. The growth factor fits its particular receptor like a key in a lock. When the right key slots into the right lock it has the same effect as turning on a switch. This is how hormones can affect some tissues and not others. An example of this is the action of the female hormone oestrogen on the breast. The breast tissue grows in response to oestrogen because breast cells carry oestrogen receptors on their surface. Other cells in the body that the oestrogen washes past are unaffected, so long as they do not have the oestrogen receptor.

Insulin-like growth factor II reacts with two kinds of receptor, the first of which was called the IGF-I receptor. When the growth factor fits into this receptor in placental cells, it results in growth, which explains its name. What is remarkable is the behaviour of the IGF-II, or type 2, receptor. When the growth factor binds into a type 2 receptor, instead of cell growth resulting, the growth factor is taken into the cell and destroyed. In other words, the type 2 receptor has an opposite effect to the type 1.

Very surprisingly, the gene for the type 2 receptor was discovered to be paternally imprinted. Only the copy coming from the mother was active. So there were a pair of genes with opposing effects, one resulting in increased placental growth, the other tending to reduce it. A complex mechanism such as this cannot be the result of chance; but what possible reason can there be for this mechanism to have evolved? The answer is found once again in the forces of intra-genomic conflict.

In this case the argument has been developed by David Haig from the Department of Plant Science in Oxford. He and his colleagues hypothesize that the more nutrients an embryo is able to extract from its mother the more likely it is to survive after birth. The extraction of nutrient will, however, be at some cost to the mother. If carried to an extreme, it may kill her or decrease her ability to carry future children.

The father of a child gains from encouraging as much growth as he can in an embryo which carries his genes. He needs to invest in the embryo at the expense of the mother because subsequent embryos carried by her might result from matings with males other than himself. This is less often the case in humans than in mice, but the principles and their evolution are the same in the two species. Moreover, if, as in mice, the embryos in a single litter can be from two or more different fathers, then the embryos with the most active growth factor will succeed at the expense of their littermates.

The type 2 receptor protects the mother, and allows her to have the maximum number of children. The imprinting of the type 2 receptor is likely to have evolved as a counter to the unbridled effect of the monstrously greedy paternal growth factor gene. We are again in the world of Thatcher and Malthus. Where there is conflict and the free market, eventually there is maximum efficiency.

You may feel that dissecting the genetics of sex is only an example of the dead hand of science falling on an exciting pastime. If so I'm sorry, but sex is after all a mechanical business. Love, one might argue on the other hand, is something completely different. Love is a thing of the spirit, irrational, sublime, safely beyond the grasp of a mechanistic science.

This too may or may not be true. At the heart of life there is the gene, cynical and self-serving, leaving nothing to chance that may perturb its progress. Because selection of a mate decides the success or failure of the gene's propagation to the next generation, the gene is likely to have a say in the process. The genes which influence mating behaviour could be more simply described as the genes for love.

In plants, mating is capricious, stochastic, and totally promiscuous. Although exceptions could easily be imagined or found, in general pollen is spread indiscriminately: the recipients of pollen are determined by the wind, or by the patterns of insect behaviour. In animals it is different, if only because animals have wider possibilities for behaviour than plants. When behaviour does not involve sitting still, the donation of pollen or sperm and its acceptance is the subject of choice. The male may choose to donate his sperm to a particular female, or she may choose, or decline, to accept it.

The question of which sex makes the choice is the subject of debate. In some species, such as chickens and deer, and some primates, a dominant male not only selects the females with which he wants to mate, but also prevents lesser males from mating with 'his' females. The dominant male must be bigger and more aggressive and perhaps more cunning than the males who rank below him. Genes which give him his advantage of size and intellect are passed on to the maximum number of children, and the species as a whole benefits, according to the 'classical' Darwinian view of evolution.

But not all species organize their group structure around a dominant male. With the rise of the women's movement in the 1970s the role of female choice in mate selection has become increasingly emphasized. The reasoning behind this view is that males produce billions of sperm, which, like the flowers' pollen, they then attempt to spread far and wide to assure the propagation of their genes. Females have to have a different strategy. They make far fewer eggs, and often they are left with the burden of raising the children. In these circumstances they are obliged to be very careful in their choice of mate. If they choose badly, they are left to care for a bundle of bad genes. Furthermore, their own genes will be disadvantaged by the company they keep. Female choice does seem the rule in many species: the male is required to display before the female, after which she will make a judgement to accept him as a mate or not.

The question then becomes, how does the female or male select for good genes? Most research on this point has been focused on female choice, perhaps because the male display tactics deployed to attract a mate are often so obvious. The tails of peacocks were considered at length by Charles Darwin. Why should evolutionary selection operate in favour of such a bizarre encumbrance? At a less florid level, female swallows plainly select their mates by the lengths of their tails. It has also been shown recently that, to a female swallow, the symmetry of the male's tail is just as important as its length. This finding has been linked to research on humans which suggests that symmetrical features are more attractive to the opposite sex than asymmetric ones. The general conclusion from these findings is that symmetry, or the length of a swallow's or a peacock's tail, is somehow an indication of a generalized genetic fitness. Birds with the most symmetric tails are likely to have the best genes, and are likely to be in the best health.

I think that this oversimplifies matters, and makes too much of a generalization. Darwin based his ideas on meticulous and exhaustive observation, and was as a consequence usually right in his conclusions. He explained that the peacock developed its tail because peahens like peacocks with big tails. In more modern terms, we might consider the genes that select for large tails out of selfishness. That genes should exist for fashion seems entirely reasonable, with the proviso that the fashion must not carry with it a significant disadvantage. In a forest full of foxes, long-tailed peacocks would soon disappear. A long tail, or an exaggerated courtship ritual, is a luxury. They can only exist in the presence of plenty. When the struggle for existence is pressing, these frivolities do not appear.

There is a temptation to use examples drawn from animals to interpret our own behaviour. The theory that human males are promiscuous by a grand design of nature has always seemed to me to carry more than a hint of wishful thinking. In the history of humanity, dominant males, although occasionally enjoying the genetic advantages of the harem, must have made little difference to the total contents of our gene pool. Conversely, even in other primates, a pattern of female choice has been difficult to find.

Meredith Small, from the University of Cornell, expected to find such a pattern when she studied female Barbary macaques in the south of France. She watched twenty females for nearly a year and observed a staggering 500 copulations. This feat has only been matched in British society by the chairman of the board of film censors. Small was surprised that the female apes apparently did not show consistent preferences for the males with whom they mated. She then reviewed all that had been written on primate mating habits. She concluded that the theory of female choice did not hold true.

It is likely that neither a simple pattern of male or female choice is the case in our own branch of the primate tree. There is enormous diversity of the patterns of social organization and sexual behaviour in our species. Such flexibility must have conferred some considerable advantage as we spread out to cover more and more of the world's surface. We may be assured that Love, be it the meeting of eyes across the room, or the gradual deepening of feeling accrued over many years, is not likely ever to be reduced to anything as simple as a DNA sequence hidden somewhere in our genome.

CHAPTER 9

COMPLEX DISEASES

In the present decade, we have really arrived in the era of the new genetics. The single gene disorders have been conquered: given the will and the money, any gene that causes a classical genetic disorder can be hunted down. The methods for this research are now established, and so the search for single gene disorders can be called a mature science. American researchers speak about 'pushing the envelope' to describe science at the frontier: the 'envelope', I believe, is the air around an aircraft that is approaching the speed of sound. At the envelope the unpredictable and the dangerous can happen. Mature science does not push at the envelope, and is safer and consequently more boring than science at the frontier. As in geographical discovery, the thrill of finding a new scientific continent lies in being the first to recognize that a new continent exists: after that all is filling in. Everyone remembers Christopher Columbus, but how many can name those who followed him in the second or third wave of ships? In a mature science everyone has a clear idea how the system works, and the results of the experiments are by and large predictable. This is not to say they are not important: sometimes they are, but the law of diminishing returns operates remorselessly.

The next frontier in the new genetics lies with the complex genetic diseases. A complex genetic disease runs in families, yet does not have a discernible pattern of inheritance. These diseases include the modern pestilences of schizophrenia and manic-depressive psychosis, diabetes, hypertension, asthma, Alzheimer's disease, arthritis and cancer. Complex genetic illnesses are common: in the United Kingdom alone there are 400,000 people with Alzheimer's disease, 1,000,000 with diabetes, 3,000,000 with asthma, and 5,000,000 with hypertension. By contrast there are 2,500 people with Huntington's disease, 3,000 with muscular dystrophy and 7,000 with cystic fibrosis.

All of us, or members of our close families, will be affected by one or more of these diseases. As well as being caused by unknown numbers of genes, they are all characterized by an interaction between the genes and the environment. Childhood diabetes may be precipitated by a chance infection with an as yet unidentified virus: without an infection by the virus at a susceptible age there is no diabetes; children born high in the mountains are unlikely to suffer asthma, since the air at altitude is dry and the house dust mite cannot survive; without cigarette smoking there would be no lung cancer, and so on.

These illnesses, which are also called 'multifactorial' or 'polygenic', pose formidable problems to gene hunters. This has been made abundantly clear by the unhappy experiences of the psychiatrists. Two psychiatric illnesses are recognized to be genetic, at least in part. The first of these is manic-depressive psychosis, the second schizophrenia.

Manic-depressive psychosis is characterized by wild swings of mood. At one extreme the unfortunate victim is extraordinarily happy, talking incessantly, joking and punning, constantly on the move, and frequently sexually hyperactive. To be with someone who is in the full florid phase of the illness is at first exciting and fun, until it becomes apparent that they are not in complete touch with reality. (A loss of contact with reality typifies a psychosis, and is usually due to a dysfunction of the 'hardware' of the brain itself. A neurosis, on the other hand, is generally considered to be all in the mind, or 'software'.) The irrepressible good humour of the manic patient can cease within days, and the sufferer is then plunged into the depths of unimaginable despair. This kind of depression, called 'endogenous' because it comes from within, is held to be the unhappiest state to which humanity is prey. One sees oneself as completely worthless, a dreadful burden to family and friends. There is no sleep and no response to kind words or to logic. Small wonder that, untreated, these depressions can end, not with a handful of sleeping tablets, but with a gun or a rope. The illness is sometimes typified by a tendency either to mania alone, or to depression, in which case it is called a 'unipolar' disease. In a milder form the condition expresses itself in deeper and more frequent mood swings than normal: people with personalities of this type are known as 'cyclothymic'.

Schizophrenia is no less awful a condition, beginning in the midteens. Before adolescents with schizophrenia become ill, they frequently have slightly odd personalities and are introverted and shy.

Because they are often 'loners' the gradual development of chaos in their minds may go unnoticed. In the florid state of the illness schizophrenics lose the ability to differentiate between the real world and the imaginary. They hear voices, which are characteristically mocking and derisive. Occasionally the voices are violent, and very occasionally a schizophrenic can become irrationally and terrifyingly violent. Schizophrenia tends to burn out, but with the dying of the flames of insanity there is also a death of the underlying personality. The chronic schizophrenic is mute and listless, unwilling and unable to care for himself.

Schizophrenia and manic-depressive psychosis are found more commonly in the families of people with the illnesses than in the general population, and are partly due, in an unknown degree, to genetic differences between those who have the illnesses and those who do not. Manic-depressive psychosis seems much more clearly genetic than schizophrenia. With both types of illness large families have been described in which the diseases descended through many generations and through many ramifications of the family tree.

It was not surprising, given the spectacular successes of genetic linkage studies of straightforward illnesses such as cystic fibrosis, that geneticists and psychiatrists should try to map the genes for mental illness. Two problems were to stand in their way, one acknowledged and the second unrecognized.

The first problem was that it is very difficult to draw a line between normal and abnormal behaviour. A person in the full throes of mania can be diagnosed by the man or woman on the street; but a cyclothymic personality merges indistinguishably into a moody adolescent. Moderate mania may be a positive advantage to a well organized and ambitious individual, who can work long hours without sleep, and can motivate others by his enthusiasm and drive.

Schizophrenia is also much more than just an illness. The bizarre ideas of the schizophrenic, if at least half in contact with reality, may contain much that is novel and exciting. Schizophrenics do not think logically: they can think by a 'knight's move', with ideas that move sideways instead of forwards. Thus in answer to the question 'Where is the dog?', a knight's move answer might be 'Yellow is a nice colour for a kennel.' Schizophrenics also suffer from 'ideas of reference' – paranoid attributions of meanings or motive to the innocuous actions of others. When you are sitting by yourself in a restaurant and the other two customers are sharing some secret joke, your feeling that they are laughing at you is a mild example of an idea of reference.

Without the paranoia, ideas of reference and knight's move thinking can also lead to completely new perceptions of the apparently mundane and ordinary; in other words, being creative or genuinely poetic. It is therefore the schizophrenic who is responsible for the commonly held view that exceptional brilliance is akin to insanity; undoubtedly, this is sometimes true.

The author James Joyce was probably schizophrenic. He had a daughter who was definitely so, and who was committed to an asylum. When he wrote *Ulysses* Joyce was mostly sane, and *Ulysses* is an undisputed work of genius. When he wrote *Finnegan's Wake*, with protracted passages in Eskimo and in his own invented language, I suspect that Joyce was not quite sufficiently in touch with the real world to control his creativity.

Ludwig Wittgenstein was a fabled Cambridge philosopher, Viennese in origin, who argued about everything with extraordinary conviction and vehemence. No-one, not even Bertrand Russell, could understand most of what he said. He was an extraordinarily eccentric individual who exhibited some signs of schizophrenia, including paranoia and knight's move thinking. Wittgenstein gave enormously popular lectures in Cambridge, perhaps because he exemplified eccentricity in a very pure form; he would, for example, stretch his hand up in front of his eyes, staring at it silently for minutes on end in the middle of a lecture.

The total opacity of much of Wittgenstein's philosophy and most of *Finnegan's Wake* is taken by experts in their respective fields to mark the highest signs of brilliant intellect. The truth is more likely to be that they are incomprehensible because their authors were barking mad. Sadly, schizophrenic individuals without the intellectual or artistic gifts of Wittgenstein or Joyce are not usually brilliantly creative. They are just unwell and often very unhappy.

Thus the very stuff of madness is, in milder doses, beneficial to human society, and the disordered thoughts of the organically mad give insights into the workings of the normal human mind. If schizophrenia is caused by defects in genes – that is, it has a physical basis – then allusive and lateral thinking is 'hard-wired' into our brains. Similarly, manic-depressive psychosis tells us that our moods are not completely under our voluntary control, that they too are part of our hereditary endowment. Identifying the genes that predispose to these illnesses and discovering their roles will not only help us understand the disease process, but might also tell us something about the construction of our mortal souls.

The problem of defining normality caused considerable difficulty during the early attempts to find the genes for psychiatric illness. This was expected. The second, unanticipated, problem was the method used to decide if linkage was real or not. Localizing a disease gene on a particular chromosome rests on finding a map marker inherited, along with the illness, in a family or several families. If a marker is inherited with a particular illness 49 times out of 50, then the disease is likely to be linked to the marker. If the proportion is only three out of four, then the association is not at all certain. Nine out of ten make it more likely, but is it proof? To make decisions about these sorts of numbers researchers must rely on statistics. Deciding the probability of genuine linkage between a disease and a map marker is far from simple, particularly in large families.

A very clever mathematician called Newton Morton apparently solved the problem. Morton is now about sixty; like many mathematicians of his generation, he dresses carefully, and I have never seen him without a beautifully knotted tie. His beard, too, is always most neatly trimmed. In his heyday, in the mid 1950s, he invented a mathematical technique for estimating linkage probabilities, which he christened the 'lod' score (lod stands for 'log of odds'). Used properly, the lod score method is an elegant means of deciding on the presence or absence of linkage, and in the years since its invention it has become the standard method for doing so. Morton himself has made other important contributions to mathematical genetics, but has also become famous for stinging attacks on other geneticists. It has even been suggested that one has not achieved anything worthwhile in genetics if Newton Morton has not attacked you.

The lod score has several quirks. Morton himself laid down a rule that the 'prior probability' of linkage had to be allowed for in interpreting the lod score. In choosing to apply prior probability to his lod score, Morton was influenced by the Reverend Thomas Bayes, another clever fellow, alas long dead. Bayes invented a theorem in which the aggregate probability of all possible outcomes is equal to one. In some circumstances this makes perfect sense: if you toss a coin a hundred times, and half the time it comes up tails, then the probability of tails is 50%. It is not hard to infer that the probability of heads is one minus a half – also 50%. Difficulties arise when this logic is applied to more open-ended circumstances, such as horse races. A person with a Bayesian outlook might follow the results of the races and note when there was an unusual run of winners without the number four, and then bet steadily on number four, because the law of averages demands that four must get an equal share of appearances, and so it must come up soon. Unfortunately this is not so: no matter what went before, at the next race four has no greater chance of winning than all the other numbers.

Bayesian logic is applied slightly differently to lod scores. Morton argued that you have to decide how likely finding linkage was when you started before you can interpret what the lod score means. Thus, although a lod score of 3 means, in its crude form, that the odds in favour of linkage are 1000:1, the fact that it is not likely that you will find linkage in the first place (*a priori*) means, Morton and his disciples argue, that a lod score of 3 does not mean odds of 1000:1, but only 20:1.

If you find this confusing, then so does everyone else. No-one, Morton and a few others apart, really understands prior probabilities. Indeed, their use in other branches of statistics is uncommon and controversial. The effect of all this is that, in the absence of a great deal of experience, it is very difficult to assess exactly what a lod score of 3 means. At the beginning of the 1980s hardly anyone had had such experience. All that most geneticists knew was that you put your data in one end of the computer program, and at the end of the program's run you were left with a number between minus infinity and plus 10 or 20. If the number was less than minus 2 Newton said you were unlikely to have linkage. If the number was greater than 3 then Newton said linkage was likely.

With the single gene disorders, when the inheritance of diseases was clear, and it was clear who was unaffected and who was affected, the lod score worked well. The threshold of 3 for holding linkage to be likely was conservative, and so few mistakes were made. With more complex illnesses the calculations of the lod score relied on several assumptions about the genetic behaviour of the disease. The mathematical terms used to describe the inheritance of a disease are known as parameters. As no-one knew how to arrive at reasonable estimations of the parameters for most of the complex diseases, it seemed that they had *carte blanche* to guess at them.

What investigators did not understand was that, by varying the parameters, a researcher could elevate or depress the lod score almost at will. Furthermore, changing the diagnosis of only one or two key individuals could produce changes in the lod score that meant changes of a hundred- or even a thousand-fold in the odds for or against linkage. The lod score is thus very difficult to use correctly and, perhaps like Newton Morton himself, it is very unforgiving of errors.

All this meant that the principal mathematical tool for deciding the validity of linkage not only gave unstable results in inexperienced hands, but that the interpretation of a given result was determined by a methodology that was imperfectly understood. In large families with mental illness there were inevitably several people who were on the borderline between normal and abnormal. In these circumstances disaster was inevitable.

At the end of 1987 a paper from the universities of Miami and Yale appeared in *Nature*. The paper was based on an Amish family in whom manic-depressive psychosis seemed to be inherited as a dominant condition. The Amish are a fundamentalist sect who intermarry frequently and, like the Mormons, have very large families. The *Nature* paper showed genetic linkage of the manic-depressive illness to a region on chromosome 11. Enormous excitement followed in the scientific press and in the general media.

A year later the British psychiatrist Hugh Gurling published another paper in *Nature*, this time showing a linkage between schizophrenia and chromosome 5. Again there was a great deal of interest in the results, but there was more scepticism about this linkage than that relating to manic-depressive psychosis reported in the paper published the previous year.

The problem with Gurling's pedigrees, it was felt, was that there were too many individuals who were on the borderline between schizophrenic and normal. There were also individuals with mental illnesses that were not obviously schizophrenic, such as alcoholism and depression. Gurling labelled these individuals 'marginal phenotypes'. In the same edition of *Nature* there was another paper, from a group led by Kenneth Kidd. Kidd was one of the Yale researchers who had reported the linkage between manic-depressive psychosis and chromosome 11 the year before. Kidd's group had now looked with chromosome 5 markers at other families with schizophrenia. He did not find linkage.

Gurling was therefore under a great deal of pressure immediately after the paper appeared, and it is understandable if he reacted with some anger to the criticism. A series of other papers appeared, each showing non-linkage of schizophrenia to chromosome 5. These papers were often of poor quality, using inappropriate analyses to disprove linkage; reading back over them now, it is plain to see that they were often written in an accusatory style. Scientific meetings of psychiatric geneticists now resembled bull-baiting sessions, rather than the rational exchanges of ideas and information that such meetings were intended to be.

Matters had deteriorated even further a year later, when Kidd's group published another paper in *Nature*. This time they were retracting their initial claim of a chromosome 11 linkage to manic-depressive psychosis. They were forced to do this because they had found other people in the original family who didn't show linkage to chromosome 11. Even worse than this, one family member, who had originally been categorized as normal, had in the meantime developed florid manic-depressive psychosis. Finally, it was also conceded that there had been some technical errors in the original DNA typing. This retraction was a remarkable display of gamesmanship, in that it received as much favourable press coverage as the original claim of linkage.

In the four years since Kidd's paper, there have been only one or two proposed linkages to a psychiatric illness, and not a single generally accepted result. Tempers have quietened, or at least bad tempers have received less publicity. The improvement in temper was probably because most of the protagonists were exhausted, and because there was a general sentiment that new approaches needed to be invented before any progress could be made.

These episodes resulted in a widespread feeling of deep suspicion concerning linkage and complex diseases. The editors of *Nature* lost their nerve completely, instructing their referees that they were not going to accept any new claims of linkage. This policy lasted for nearly two years but, prestigious as *Nature* was, it was not able to hold back the gene hunters. Psychiatric illnesses were too difficult and too dangerous, but there were other prizes.

It was at the beginning of the century that a German neurologist, Alois Alzheimer, became interested in senile dementia. Before Alzheimer it had been assumed that senile dementia, the loss of mental capacity that affects the elderly, was a normal state, and that perhaps anyone who lived long enough would eventually end up demented. Alzheimer, however, set himself the task of studying the brains of people who had died with dementia. Under his microscope he could see tangles of dead and dying nerve cells that were absent from the brains of elderly people without dementia. These tangles were focused on dense congregations of unknown composition which he called senile plaques. The plaques were also absent or rare in the brains of other elderly people. Senile dementia was indeed a disease.

As our population has grown more healthy and lived longer it has become clearer that dementia is not inevitable: most people who live to a ripe age retain most of their intellect. However, as the average age of the population has risen, the number of people potentially incapacitated by dementia has increased enormously. Today, the disease which bears Alzheimer's name affects a million Britons and four million Americans.

Before the new genetics, the problem of establishing the cause of Alzheimer's disease had proved impervious to the techniques of modern science. The senile plaques seen in the brain at *post mortem* were made up of a strange substance called beta-amyloid. No-one knew exactly what beta-amyloid was, or why is was present in the plaques of Alzheimer victims. In the absence of any new leads, and because Alzheimer's disease sometimes seemed to run in families, the illness became a target for positional cloning.

A key observation preceded the search for the Alzheimer's gene. Neuronal tangles and plaques had been found in the brains of adults suffering from Down's syndrome, which is caused by the presence, from birth, of an extra copy of chromosome 21. It was reasoned that this apparent coincidence might mean that a gene or genes on the same chromosome was implicated in Alzheimer's dementia.

Collecting families with the disease proved difficult, because the diagnosis can only be definitively made at autopsy. Nevertheless, in 1987, St George-Hyslop (who with a name such as his could only be English) tested chromosome 21 markers in a few families. These families suffered particularly severe dementia which came on early in life. The early onset was helpful for the researchers because it made it simpler to sort out who had the disease and who did not. When St George-Hyslop and his group analyzed their data, they found weak evidence pointing to the presence of the gene causing Alzheimer's disease on chromosome 21.

Other investigators tried the chromosome 21 map markers. Some thought that they might have found linkage, and others thought that they probably had not. In retrospect, the weakness of the evidence for and against linkage was due to the various investigators acting responsibly: in other words, they were not tweaking the statistics one way or another to suit their particular prejudices or grant applications. Also, there was a twist to the story that probably held back potentially vociferous critics. At about the same time as St George-Hyslop's chromosome 21 mapping result was published, the gene for beta-amyloid had been examined by Rudolph Tanzi at the Massachusetts General Hospital in Boston. He found that it arose on chromosome 21, near to the point mapped for Alzheimer's disease by St George-Hyslop.

Although the gene for beta-amyloid was close to where an

Alzheimer's gene should have been, the proximity was not quite sufficient to be sure, one way or another, if beta-amyloid and Alzheimer's disease were functions of the same gene. There followed a period of protracted argument about whether linkage existed at all, or whether there was more than one gene causing the syndrome, or whether the disease was even genetic. Although very worrying for all concerned, the conduct of the conflict was quite different from the raging quarrels that surrounded the genetics of schizophrenia and manic-depressive psychosis. This can be explained by character differences between different types of medical specialists.

It is a generally accepted wisdom that a cardiac surgeon is never happier than when he has saved a patient from the nearest of near-death experiences with a bloody and dangerous operation. Stereotype or not, the observation is, in my experience, completely correct. (Surgeons, incidentally, are often excellent molecular biologists, because they have the effrontery to confront the difficulties of arcane molecular 'cookery' head-on.) Neurology and psychiatry attract equally polarized character types. Psychiatrists are attracted to their speciality, it is said, because they are driven to come to grips with their own quirks of personality; emotional by disposition, they are then trained to be deeply suspicious and manipulative of all human nature. A psychiatrist examines his patients by talking to them, taking the broad sweep of their thought and mood as diagnostic of underlying pathology. Neurologists are completely different. They are worriers, full of fret and bother – in psychiatric terms, thoroughly anally fixated. A neurologist keeps in his doctor's case a set of needles and hammers and tuning forks and odd-smelling little bottles. The neurological examination is an obsessive ritual, beginning with the 'higher functions' of memory and calculation, descending remorselessly through every nook and cranny of the nervous system, and finally concluding with the sense of position in the toes.

Sadly, this contrast in styles means that neurology clinics are full of people who are mad rather than organically unwell, and that psychiatrists inevitably have under their care numbers of patients with intractable and unrecognized physical illness.

To a neurologist, then, even one of the type who goes in for gene hunting, the sort of uncertainty that surrounded Alzheimer's disease was profoundly unsettling. The best analysis that anyone could come up with, carried out by St George-Hyslop on most of the families from all groups, suggested that more than one gene was likely to cause the illness, and that the chromosome 21 linkage was active only in families in which the disease began early in life. The chromosome 21 linkage remained undecided, leaving everyone in the field in a state of limbo.

Suddenly, in 1991, Alison Goate and others from St Mary's Hospital in London found a mutation in the beta-amyloid gene. The publication of this result was greeted with universal scepticism, followed, within a month or so, by grudging acceptance, and then, within a year, by a flood of papers showing how the mutation could predispose to the tangles. Alzheimer's had started to give up its secrets. The four years in limbo had been rewarded by a translation to scientific heaven. The Mary's team were head-hunted by the United States, and left almost *en masse* for Florida. Florida is the chosen place of retirement for many of the very richest of Americans; if money could buy a cure for ageing, then the cure was likely to be found in Florida.

In the same year another Alzheimer's linkage was reported, this time on chromosome 19. The linkage came from Duke University, in North Carolina in the USA, and was statistically significant only if a new method of analysis was applied to the family data. This technique dealt with uncertainties of diagnosis by exclusively counting definitely affected people. It was invented by a geneticist called Dan Weeks.

Scientists are highly suspicious of any analysis carried out on a subset of data, because of the fear that the data used has been selected in order to hide the weak points. As Dan Week's method was new and untried, its sound theoretical basis and the chromosome 19 result were ignored by most other groups studying the illness.

However, also at Duke University were two biochemists, Warren Strittmatter and Guy Salvesen, who were using a non-genetic method to find clues to the nature of Alzheimer's disease. Late in 1992 they took the beta-amyloid protein and incubated it in cerebro-spinal fluid, the liquid that washes round the brain and spinal cord. After the incubation, they found that several proteins from the cerebro-spinal fluid had stuck to the amyloid.

One protein was well known, and is called alipoprotein E (ApoE). ApoE carries cholesterol around the blood stream. Now cholesterol and ApoE might be interesting to cardiologists, but they were not substances that meant much to neurologists. Salvesen and Strittmatter's clever observation might well have remained buried somewhere in an obscure publication and never come to anything, except for two coincidences. Firstly, someone had already localized the gene for ApoE to chromosome 19. Secondly, Strittmatter and Salvesen were at Duke, the same institution as the geneticists who thought there was an Alzheimer's gene on chromosome 19.

Thus, at least within the walls of Duke University, the association between ApoE and Alzheimer's became wonderfully interesting. ApoE is a polymorphic gene, that is it comes in several different types (see page 31 ff). This was a good first step. A gene clearly has to be variable if it is to make some people more susceptible to disease than others. There are three common types of ApoE gene, at least in those of us whose genes are of European origin: ApoE2, ApoE3 and ApoE4. In the population at large, 90% have one copy of ApoE2, the most common variant, and about 30% have one copy of ApoE4. The Duke scientists looked at the variants of ApoE in thirty people with Alzheimer's disease and the proportion of those with the ApoE4 was about half, 20% more than in the normal population. These results prompted a bigger study from Boston, this time of 500 people with Alzheimer's disease. The proportion of ApoE4 was 64%, twice that expected. Finally another study from Duke was reported in *Science*, less than a year after Strittmatter and Salvesen's first results. The Duke scientists showed that the risk of Alzheimer's disease at age 75 was 20% if ApoE4 was not present; this figure rose to 45% if one copy was present, and to 90% for those unfortunate people with two copies of the gene.

As if two genes for Alzheimer's disease were not enough, the ApoE results overshadowed another important result. A few weeks before the Duke University's report, another paper in *Science* reported a third gene for Alzheimer's, this time on chromosome 14. 21, 19

These experiments have led to an extraordinary explosion of knowledge about Alzheimer's disease. Five years ago no-one would have disputed the fact that its cause was a complete mystery. Now the risk of developing it can be assessed on a large scale, and for the first time targets for therapy can be seen.

This success has been possible because, when faced with the problems of uncertainty about linkage, the various research groups did not simply decry each other's results. Possibly because many of them were obsessive neurologists, they instead worried about their own findings and analyses, and continued to collect families on the massive scale that was necessary to make sense of the problem. They were also, like many good scientists, not a little lucky. Finding beta-amyloid on chromosome 21 was to some extent deserved, but ApoE4 on chromosome 19 can only be seen as a gift from above.

The experience with Alzheimer's disease shows that reverse genetics are effective with complex problems. The statistical problems of deciding if linkage is real or not can be solved by throwing data at them, in other words by collecting hundreds of families instead of a hundred subjects in just two or three families. The fairy godmother factor, finding a 'candidate gene' like ApoE4, already discovered and waiting for you on the chromosome of your choice, is a little more difficult to control. So too is the element of luck in finding linkage before you are exhausted by the struggle from chromosome 1 to chromosome Y.

Scientific research works much more effectively, and is certainly less nerve-wracking, if you don't have to rely on luck. The way forward was first shown by John Todd, who has his laboratory one floor below mine in Oxford. Todd wants to find the genes that cause diabetes, and he is a man in a hurry. He has a Northern Irish accent that cuts like a hot knife through the butter of English thought. Todd and I both swim for exercise at lunch time, to clear the cobwebs for the rest of the day. Todd churns up and down the lanes like a shark in search of a square meal, while I do my impression of a whale in shallow water. Afterwards in the shower, stark naked and dripping wet, Toddy harangues me about the finer points of genetics.

Todd was studying an animal model of diabetes, a type of mouse called the NOD mouse which spontaneously develops a genetic form of diabetes. He was doing this because it is potentially much easier to map genes in mice than in human beings. Laboratory strains of mice are completely inbred, that is, within a given strain, they are all genetically identical. By breeding between two strains it is possible to generate hundreds of offspring from the equivalent of two parents. This is obviously an improvement on the human condition, where the number of children is limited, where paternity is never entirely certain, and where permission to carry out a study has to be negotiated in advance.

Before John started his work on the NOD mouse, many mouse genes had been mapped, but most of the mapping had been done by crossing and intercrossing mice with various defects or traits deriving from known chromosomes. This is an extremely slow and laborious way of working. What was needed, he recognized, was a set of mouse DNA markers to match the genetic map in humans. He concentrated on microsatellite repeats, regions of DNA where simple sequences such as CACA are repeated twenty or so times. The length of the repeated sequence varies greatly between strains of mice. Microsatellite repeats are very common, and many published sequences from mouse genes were known to contain them. Todd and his team thrashed through mouse sequences to find as many microsatellites as they could, as well as finding new repeats for themselves. Within two years Todd had covered most of the mouse genome with the repeats, and had found the chromosomal location of four genes that caused diabetes in NOD mice. Subsequently he has found several others. The race is on to test if the same genes cause the human disease.

This is to me an amazing piece of work: John had realized that a mouse marker map was necessary, saw that this could be accomplished by using microsatellites, and drove himself and his team until the map was finished. I have heard him berating the people in his lab; 'You must think of yourselves as master craftsmen. Your Gilsons are the tools to make your works of art!' (A Gilson is the standard instrument for measuring out the minuscule volumes that are the ingredients of molecular biology.) A more faint-hearted individual than Toddy would have only seen the difficulties, and might never have started.

Following the same lines as Todd, Mark Lathrop in Paris could map genes for hypertension in a hypertensive rat. Lathrop had learned his genetics from Ray White in Salt Lake City. His fame previously rested on his computer programs for estimating lod scores: almost all geneticists now use Lathrop's linkage program. Lathrop found two genes; one of these immediately suggested a human counterpart and this has already been shown to have a role in human illness. Lathrop's and Todd's studies had shown for the first time that it was possible to take a complex trait and go systematically through all the chromosomes until all or most of the genes predisposing to the illness had been localized.

Mark Lathrop then worked with Jean Weissenbach at the Généthon in Paris to apply the microsatellite technology to the human map. The result of this effort was an extraordinary tally of 850 markers covering all chromosomes. These markers were ideal for mapping genetic diseases. The title of the paper in *Nature* was 'A second generation linkage map of the human genome'. The future had arrived. Even better than the sheer number of the markers was the fact that the technology used to type them leant itself to automation on a large scale.

Starting at the beginning of the first chromosome and going through to the end of the last is called a complete genome search. Before Todd's and Lathrop's papers, gene hunters had stopped searching as soon as they had found linkage. This was partly because the diseases they had been studying were mostly caused by one gene, but was also because it was just too difficult to search the whole genome.

Now a complete genome search could become routine, because it could be completed before everyone in the lab died of boredom and frustration. Gene hunters had progressed from the stage at which they were like nineteenth century prospectors, scratching round for a single lucky strike, to conducting operations comparable in scale to a geological survey of an entire continent.

The way forwards towards finding linkages for all the other complex genetic traits is now clear. First, collect many families, perhaps amounting to a thousand individuals, then, as the White King said to Alice, 'Begin at the beginning, go through to the end, then stop.' This is expensive: to pay someone to recruit a family of four and to study the family in detail can cost nearly a thousand pounds. To carry out a complete genome search on the scale of hundreds of families will cost up to a million pounds. But this expense needs to be taken in context. The cost of a genome search for hypertension, when divided by the five million people in the UK with the illness, is about 20p per head. The cost to a pharmaceutical company of bringing a new drug onto the market is £100 million. Compared to this, a genome search is a trivial exercise.

CHAPTER 10

ASTHMA

So, what is it like to be a scientist? Is it to live in an exalted world of high intellect and cold planning, with each step calculated to carry you forward on the road to the inevitable right answer? Hardly so. My research into the genetics of asthma has taught me that the process of science is not at all a simple matter. Science goes forward by wrong assumptions and half truths, and only in the end is everything clear.

My interest in the genetics of asthma began when Julian Mergwlin Hopkin came to Oxford. He knew a bit about genetics and had a mad Celtic idea that a gene might cause asthma, and that it would be a good idea to look for it. Julian's idea caught my imagination because it fulfilled the first rule of science: ask the biggest question that you can.

That summer I had heard a molecular geneticist called Ron Wise talk about new discoveries concerning oncogenes, the genes which cause cancer. His research seemed to me quite beautiful; beautiful because it was fundamental, explaining what cancer really was. I had decided there and then that I wanted to get into genetics if I could. Here in Oxford, Steve Reeders, apparently with minimal effort, had just found that the gene for adult polycystic kidney disease was on chromosome 16, and finding genes was obviously a doddle. I offered to help Julian, and so it was that Hopkin and Cookson set out to conquer asthma.

Having decided that we were going to do it, we practically had to be restrained from starting to collect families the very next day. In fact we spent about six months planning what to do. A first step was to read all previous scientific papers on the genetics of asthma. Earlier pundits on asthma included Maimonides, who had recognized it to be familial in the twelfth century, and it didn't seem to me that things had moved on greatly since then. Amongst the more modern authors there

was some sort of general agreement that there was a genetic component acting somewhere in asthma, but no-one was sure whether there
was one gene involved or many, or how much non-genetic factors
influenced the appearance of the illness. I remember thinking that it
was obvious that no-one had studied it properly, and I never thought
that we would fail to do a better job.

The lack of self-doubt in the face of one's own crushing ignorance
is, it seems to me, a critical component of original scientific thought:
you have to believe that you can improve on everything that has gone
before, and you cannot be afraid to attack established orthodoxy.
Unfortunately, the obverse of this desire to throw away anything done
by anyone else preceding you is that the world is full of people who
have ignored orthodoxy, and by and large they are failures.

The planning that Hopkin and I did involved deciding how we
would examine the families we were going to recruit into the study:
what questions we would ask them and what tests we should do on
them. I had driven round the Australian Bush testing for asthma and
allergies in seasonal grain workers, and I had a reasonable idea how it
should be done in families. The philosophy thrashed into me by my
boss in Perth, Bill Musk, was that you should try and get as much
information as possible from your subjects, and then go back and get
some more. I didn't know anything at all about genetics, but Reeders
and Kay Davies explained that if I bled the subjects white and stuffed
the samples in a cold enough freezer, the DNA would wait until I was
ready for it.

Thus, with minimal experience and few preconceptions, we started
searching for families. This is the rough equivalent of loading up a
canoe with a bundle of provisions and setting out to explore Africa. I
knew that if we sat in the hospital and asked the families to come to us
they would not do it. After all, who would volunteer to visit a doctor
they had never met, in order for him to do several highly uncomfortable tests for reasons not possible to understand? I therefore used to go
and visit the families at home, mostly in the evenings. I would load up
with dozens of syringes and needles and little bottles of house dust
and pollen extracts and a vitalograph for measuring lung function.
The 'portable' vitalograph weighed about sixty pounds, which made it
feel even more like exploring Africa. Miniature electronic machines
existed, made by the same company, but I believed the mechanical
device gave more reproducible results, and I insisted on taking it.

I would usually test all the members of the family that I could. The
received wisdom was that large families would be the best ones to

study, and so I doggedly traced uncles and aunts and first and second and third cousins all over England. After the journey it took a minimum of thirty minutes to carry out all the tests on each person, or two and a half hours per household. Testing was invariably conducted with other family members looking on with undisguised apprehension. The process was quite tiring and, having tested about 300 people in this way, I couldn't go on. This aspect of the work was continued by Pam Sharp, a research sister who tested another 900 subjects without ever flagging in her enthusiasm or her ability to charm an armful or two of blood out of a recalcitrant father. (It was always the men who refused the tests.)

Anyone who is unsure of the value of the human race should carry out such an exercise. I never failed to be surprised at how good-natured and helpful most family members were. They would meet me at the station and drive me to their house, offer to feed me, and hardly ever complain when I stuck needles in them and made them wheeze. This was especially remarkable because most of them didn't consider themselves unwell. It is understandable that parents of a sick child would want to help, but many of the people I was testing were not ill, even if they had some minor symptoms, and their willingness to help was motivated purely by kindness and altruism.

As I worked my way through the families, I became convinced that we were studying an authentic genetic illness. Asthma is part of a syndrome called 'atopy', which is pseudo-Greek for 'unclassifiable disease'. People who have atopy tend to be allergic to things like house dust and pollen, and get eczema and hay fever besides asthma. Julian and I reasoned that we needed to test for atopy, and not just for asthma. Most of the time it was easy to tell who had atopy and who did not. When one parent was affected then half of their children seemed to suffer. As a medical student I had learned about the thalassaemias, which are illnesses caused by errors in the genes for haemoglobin. In my textbooks had been diagrams of the abnormal genes, and how they had combined with normal genes to produce different kinds of thalassaemia. I imagined that I could see the same sort of genetic abnormalities segregating the families I was examining.

We were concerned about the pattern of inheritance of the illness, that is to say whether it was dominant or recessive. We thought that atopy was a dominant illness, because we didn't see carriers very often. Some years later we were to see that this was a hopeless over-simplification, and not even half right. Our results were greeted in the UK with universal indifference. On the other side of the Atlantic, where much of the earlier research had taken place, the sentiments induced in some quarters were less sanguine.

We persuaded the Professor that we had enough evidence to show there might be a gene, and enough families in the freezer to start looking for it. He said yes, except that there was no lab space free in the department. Sir David Weatherall had built the Department of Medicine up from scratch over fifteen years, and was such an effective collector of scientists that they were tripping over each other in the corridors. Some laboratories, particularly the muscular dystrophy group, were so tightly packed they would turn on strangers and start biting them and each other like overcrowded rats. Even the cupboards seemed to be full of molecular biologists.

Under these circumstances I could only learn molecular biology at nights, and at weekends when there was some free space, as the number of people working would drop by a half. I was taught what to do by the extraordinarily sarcastic Richard Wells. Dick was a Rhodes Scholar. As well as being very bright – 'You look very smart today Dick.' 'Thank you. I am smart.' – Dick was also phenomenally hard working, so whenever I went to work Dick was already there. The amazing thing was how many people were always there; molecular biology seemed custom-made for workaholics. Dick taught me the basics and on Saturdays and Sundays and weekdays after six we would toil away extracting DNA from my blood samples.

Eventually Mark Gardiner, the Reader in Paediatrics, kindly let me share his laboratory, so that I could work on weekdays as well as at other times. There were two Marys in Mark's lab: Mary D and Mary K. They both knew a great deal more about lab work than I did, and in addition Mary K knew a great many more bad words. Only when she had explained that she had had a convent education did I understand how such an angelic face could conceal such a colourful tongue.

Also in the lab was Jo. Jo is quintessentially Oxbridge, straight-backed, fair-haired, clear-eyed, speaking in rounded vowels, and frighteningly clever. Anyone in the world would instantly recognize her origins. The first time we met she greeted me rather as Lady Bracknell might offer pleasantries to a railway porter: 'Who are you? What do you do?' Unlike Lady Bracknell, Jo has a heart of gold. She also has some rather endearing eccentricities. Once, I recall, she was too involved in her experiments to notice that the ladder in her tights had formed a complete circle. She continued bravely, ignoring the severed nylon, draped around her ankle like a schoolboy's sock. Over the next year we sparred incessantly, she giving rather better than she got.

In this environment, I began the grand genome search for the asthma gene. Very early on I began to realize how inadequate the families I had collected were. The ability to map a disease to somewhere in the genome depends critically upon how sure you can be whether family members have the disease or not. 'Marginal phenotypes' was a term coined by the psychiatrist Hugh Gurling, who used it to describe subjects who could be normal or abnormal, depending on how you looked at it (see page 114). In a genetic linkage study, marginal phenotypes are asking for trouble because, unless you are very careful, they can be used to push a result one way or another. When I looked at the largest family we had, and excluded all the uncertain diagnoses, I realized that I could use less than a quarter of the family members we had tested for the linkage study.

My distress at this discovery was matched by my distress at my technical inabilities in the lab. What I was trying to do was, in molecular genetic terms, very simple. I had to cut DNA from each of the samples with restriction enzymes and test them with radioactive markers from different chromosomes. The results were seen by exposing X-ray film in a freezer overnight at —80°C. In the morning I would risk frostbite to take the cassette containing the X-ray film out of the freezer and rush it downstairs to be developed before condensation dampened the film. In the darkroom I would struggle with the frozen cassette to remove the film and place it between the rollers of the developing machine. The rollers made me understand why real scientists do not wear ties. After the film was in the machine I could leave the darkroom and wait outside for the developed film to appear.

What I always hoped to see was a nice neat pattern of bands, like a simplified bar code on goods at the supermarket. In the first months, what I usually saw more resembled aerial photographs of a B52 nocturnal bombing strike on Baghdad. By way of variation, the bombing patterns were sometimes replaced by impressions of clouds by moonlight, or at other times someone's skin after smallpox, or, most distressingly of all, a complete blank. Mary K and Mary D were very kind to me during this period, but for a while I began to doubt whether what I was doing would ever come to anything.

I started with chromosome 13, because someone years ago had thought there might be linkage with atopy and a marker there. It took me three months to realize that the chromosome had nothing to do with atopy. One down, less than 1% of the genome excluded; if I had studied enough families, which I seriously doubted, 150 probes would cover the genome: 150 times three months, or 37 years and a bit.

At this stage I realized that molecular genetics was dangerous for your health. I was, and remain, lazy by nature, working only when I had to. If there was a lot to do, then I would work flat out so that I could finish the job and relax again. This policy had stood me in reasonable stead over the years, so that I could spend much of my time in a desirable state of relaxation. Now, for the first time, even a sustained burst of genuine work was going to be nowhere near sufficient.

I worked as hard as I could. We were thrown out of our house because our landlord's wife wanted to sell and buy a flat in London. At some stage I may have had a beard, but no-one noticed. I slept badly, and developed the symptoms of a pyloric ulcer. The autorads began to look a bit better, and sometimes I could type in the results and get a negative score from the fiendishly difficult linkage computer programs. Over a year I struggled through 16 probes. 10% of the genome, ten years to go. The bank manager was polite, but distrustful.

And then Babs in Kay Davies' lab at St Mary's Hospital told me about the Jeffrey's probes. These chromosome markers were derived from his genetic fingerprint research (see page 37). Babs said they were OK. I wrote to ICI, who owned the patent, and asked for permission to use them. They gave me permission to use five of the probes. Rather than pay ICI's handling fee, I persuaded Babs to give me the box of probes that she already had from them.

In the box were seven probes.

'Pick a number' I said to Mary D, and read out their identification numbers.

'51,' she said.

'Are you sure?' It didn't sound right to me. 51 came from chromosome 11. It was not one of the five probes I was allowed to use: Babs had forgotten to take it out of the box.

'Definitely. You'll see.' 51 still didn't sound right, but Mary D had taught me everything I knew, so I didn't want to offend her. I set up all the reactions. Everything went perfectly. Three days later I came up from the X-ray department with the autorads.

'What have you got there?' said Jo, implying from her tone that I had with me something that even a cat would not have dragged in.

'The asthma gene, of course,' I said, laying the autorads out on the viewing box so that I could read them. Jo looked at me pityingly, and went back to the lab. I put a copy of a family tree next to the autorad so that I could see who was supposed to be whose father, and who had the disease and who did not. There were three children with asthma, which they had inherited from their mother.

'Look,' I called to Jo through the door, 'these children have all got the same band as their mother.' This could easily have happened by chance. I was bluffing. However, the affected mother had an affected brother, who also had the band, and an unaffected sister who did not. There were nine in the family, and seven times the band went with the illness. Twice it went the wrong way. Two out of nine, it should have been four or five out of nine. Interesting, but not convincing. I put a second autorad up on the box, from a larger family. There were sixteen individuals, and only twice more did the band go the wrong way. Four out of 27. I did not believe it.

'Well,' said Jo, poking her head round the corner, 'What's happening?'

'Look,' I said, and we both looked again, and then Mary D came in and I showed her too. We agreed that this could not be a chance association, but I still did not believe it.

The larger family had other members whose DNA I hadn't tested, so I took the samples from the freezer and set up the reactions. I drove home that evening with my head ringing with the enormity of it all. There was a gene for asthma, and it was on chromosome 11, and I was the first person in the world to know. Or was I? Surely I had done something wrong. I told Fiona and her father all about it, and they were as excited as I was. I slept badly, going over it all again and again: was it real or not? The next morning was a Saturday, and I went back to the lab and looked at it all again before I rang Julian. He came in, and we checked through it all together. It all held up. On the computer I worked out the lod score, which measures the probability that a linkage is genuine. The score was more than three, supposedly proof of linkage. But could we believe it? Neither of us was sure. The recent experiences of the psychiatrists loomed always in the background. The acute sense of uncertainty, which in retrospect was entirely rational, persisted for the next four years.

By Monday I could read the autorads with the results for the other members of the large family. Six more individuals, five with the right band, one without. Transiently I was sure that the result was real. Julian arranged with Weatherall that we should show the results to some experienced geneticists in the department. Half an hour before the meeting I was preparing the autorads for presentation, checking the band pattern against the family trees. I came to the critical family. To my absolute horror the band pattern did not coincide with atopy in the pedigree. I had got it wrong. The assembled might of the Oxford Genetics Faculty was waiting impatiently and I was already late. I have never known more acute panic. It took five minutes of complete misery before I realized that the autorad was back to front.

Over the next three months we checked everything as carefully as we could. Other families I processed added little evidence for or against linkage. In the critical families only three or four of the phenotypes could be described as marginal. If we left them out the lod score was still above four. When the pressure began to build from the opposition I worried again and again about those marginal subjects, because they introduced uncertainty into the results. Other investigators had been trapped by false linkages; why should we expect to be different?

John Bell, now Weatherall's successor, suggested I show the data to Mark Lathrop in Paris, who had written the computer program called 'Linkage' that is universally used for calculating lod scores. The number of people in the world who could have written the program could be counted on the fingers of one hand. I did not know if I had analysed the data correctly, so it was with trepidation that I caught the early plane to Paris. Lathrop patiently went through the analysis with me. I had not made substantial errors, which was a huge relief. He was at pains to show me how sensitive the program was to 'the model', the assumptions about the probable relationship between the gene and the disease. The sensitivity, or instability, of the lod score meant that the strength of the linkage between atopy and the chromosome 11 marker could only be judged approximately.

Over the next four years I made the dawn trip to Paris many times, on each occasion learning more about genetics, both theoretical and practical. Mark was often exhausted, as he commuted between his collaborators around the world, and we would look at each other whey-faced beside the computer terminal. Without his help, I'm sure that our progress would have been minimal.

In the new year we published our results. We were given a generous grant from the Wellcome Trust to find the chromosome 11 gene. All that we could say from the first result was that the gene was within 30 million bases of DNA, somewhere near the centre of chromosome 11, on its long arm. We could set up a DNA lab of our own, and I was given a Wellcome Fellowship to pay my salary. The Wellcome Trust is a great British institution. It is the biggest medical charity in the world, bigger even than the Howard Hughes Foundation in the United States. We were told that they recognized that our research was a 'flyer', that is high risk, but that they were nevertheless willing to support us.

The second consequence of our success was that a previous grant, from the National Asthma Campaign, was not renewed. This grant paid Pam's salary, and its loss was significant. The reason for non-renewal was, we were told, because of 'the controversy' surrounding our results. We knew nothing of controversy. We had an interesting result, and we would try to repeat it. If we were wrong, and most investigators make mistakes at some stage in their careers, then we were wrong, and we would have to learn from the experience.

The source of the controversy is not important. Science is like any other human pastime, in that success, or even pseudo-success, results in enemies you didn't even know existed. Nevertheless I am sure that David Marsh would not mind being mentioned. David had studied the genetics of allergy for years, and had found a gene that controlled allergy to ragweed. He has strong opinions about many things, and expresses them forcibly. I have enjoyed his company on many occasions, and he has been a good host to me when I have visited him in Baltimore. He had been vitriolic about our results, particularly about how we had defined atopy, and it was useful for our new 'enemies' to take his opinion at its face value.

We were joined by Rob Young, a six-foot-six New Zealander with blond hair and blue eyes who, when he was not rowing for Oxford University, or windsurfing off the Turkish coast, collected a further 64 families. He was careful only to study young families, who could be more reliably phenotyped. The result was a lod score of 3.8. There was one little quirk in the data: the genetic distances between atopy and the 51 probe were greater when the disease came from the father's side than when it came from the mother's. These differences are common, but it is nearly always the female side that has the greater distance. We could not explain this, so we forgot it. The important thing was that the lod score was positive. For a while at least we could relax; it seemed that the original result was correct.

We submitted a paper describing the results to *Nature*. The paper really only confirmed the earlier results, and we were surprised that it went to the reviewers. There were four anonymous reviews. Two raised difficulties with the paper, which were soluble. The other two attacked our work. One was obviously David Marsh, who maintained his usual line of assault on our classification of the disease. A fourth review concluded:

> We must suppose that Lathrop correctly analysed the data presented
> to him. This leaves a disjunction between the incredible and the

unthinkable: either the authors have identified a gene present in 36% of the population, which their unstated classification rule recognizes with nearly perfect sensitivity and specificity in seven-generation pedigrees, or atopy was not scored independently of the minisatellite marker.

In other words the reviewer did not believe our results, and the only possible conclusion was 'the unthinkable': we must have made it up. The review was not signed, and contained a touch of mendacity, as the study was specifically of two-generation families. Inevitably the paper was rejected. More than that, we had been branded as cheats, and there was no redress against our anonymous assassin.

Comment by anonymous referees is known as peer review. Most often peer review works very well. Papers that are unsuitable are weeded out, and the referees' criticisms often improve the final quality of the paper. However, referees can hold papers back while they copy the research in their own laboratories, or they can dismiss the paper because they feel the results are incorrect, or they can pass a flawed paper if the results conform to their existing prejudice. The only check on the abuse of the power of a referee is the judgement of the editor of the journal.

Although we were reasonably sure of our results, to be called liars in this way was distressing. I realized only too well, however, that people in our position had deceived themselves about their research. Although I could not see how we might be completely wrong, it was all most uncomfortable.

Julian was then contacted by a Dr Shirakawa from Japan. Taro Shirakawa wanted to study the genetics of asthma, and to come to Oxford to do so. At first Julian said no. Taro, however, had an uncommonly persistent streak, for which we would eventually be very grateful, and kept writing until Julian said yes. A month before he arrived, he rang Julian to say he had tested some Japanese families for linkage.

'What do you think his lod score was?' Julian asked me.

It had to be bad news, 'Minus twenty?'

'No,' he said, 'It was five.'

Good news after all. We must have been right. Perhaps.

Pam had continued to recruit more families. She had advertised in *Asthma News* and in other places, and nearly a thousand families had contacted her. She had set to work with a vengeance, and most mornings there would be tubes of blood, charmed out of the populace and waiting for processing on the laboratory bench. The results of the DNA testing from the families should have been causing the lod score to go up and up, and we should have been able to settle in comfortably for the long haul to the gene itself. Unfortunately, the lod score was not going up and up; quite the opposite, it was gradually going down. We had continued to deal only in phenotypic certainties, so we could not blame that for the negative results.

The most striking thing about these families was that everyone seemed to have the disease: fathers, mothers, children and grandchildren were all damnably atopic. We could reason that this was the result of the way we had selected the families: families were much more likely to volunteer to be guinea pigs for clinical research if everyone in the family was affected.

Things were also not going well in the lab, which was my fault. The Wellcome Trust had given us enough money for salaries. I had little experience in molecular biology, and I was dealing with issues beyond the limit of my understanding. Instead of organizing the laboratory properly, I became intolerant and irascible, and at the end of an increasingly unhappy year the two post-docs left.

I made more trips to Paris to go over the data with Mark Lathrop. I had started looking at the data with a form of analysis simpler than lod scores, called 'affected sib-pair' methods. These are less 'efficient' than lod score methods, which means that many families need to be studied. However, because they concentrate on affected individuals, they are much less prone to misinterpretation and error than lod scores. With them I could show that the evidence for linkage was not due to a quirk of the way we had defined the illness.

Mark had written a little program that did the sib-pair linkage analysis. At one point he had modified the program to show differences between linkages coming from fathers and mothers, because another study, of diabetes, required such an analysis. We ran the program on the asthma data because it still gave the standard analysis in which we were interested, although the bottom four lines of output were extra.

The curious thing was that the extra four lines showed that all the inheritance of atopy on chromosome 11 was coming from the mothers' side. Mark pointed it out to me, but I didn't really understand what it meant. I gave some some sort of facile explanation which he did not particularly like, but in the absence of any better ideas we moved on with the analysis.

It was only when I got back to Oxford that I began to look at the result more closely. I thought it must be a chance phenomenon, something to do with genetic distance being higher in males in that particular part of the genome. I ran the analysis with all the other markers we had tested, and they showed the same thing. All the inheritance was from the mothers' side. What was going on? It did not make sense.

Then I remembered a conference to which I had been in Zurich. The sun had shone brilliantly on that beautiful city, and the Alps were majestic above the clouds on the far side of the lake. In the park, the junkies flapped round their pushers like crows around a piece of meat, rushing off with their dose and mainlining happily in the full view of passers-by. Rob and I had presented data at a session in which Julian was the chairman. The other three presenters and the co-chair made up the total of seven people in the room. One presentation had shown that eczema was more common in the children of allergic mothers than in the children of allergic fathers. At the time it had seemed unimportant; now it might be relevant.

After I had traced the authors of the paper to London, they were able to tell me that there was a long-standing literature on the phenomenon. Even the old generation of allergists used, apparently, to say 'the children of asthmatic mothers' as they chattered after lunch at the Royal Society of Medicine and waited for the chauffeur to arrive with the custard-coloured Rolls-Royce. I found the old papers, and there was no doubt about it: atopy was predominately inherited from the mother's side. Modern papers, studies of thousands of children, had shown the same thing.

The reasons for the maternal inheritance were unknown. It might be that a mother's immune system could affect her child in some way, programming its immunity across the placenta or through her breast milk. Another alternative was that the gene was paternally imprinted, turned off when inherited from the father. The mechanisms remain unknown. It is also obvious that some allergies can be inherited from the father, when they are due to genes on other chromosomes. Nevertheless, from that point on everything about genetic linkage in our families began to make more sense.

For the first time I became wholly confident that the linkage to chromosome 11 was real. Instead of conjuring a lod score out of a computer by making a series of critical assumptions about unknown things, the affected sib-pair analysis gave simple counts of chromosome-sharing which everyone could understand and interpret. Andy Sandford and Miriam Moffatt, who were graduates studying for their PhDs, began to take control in the lab. As they went from strength to strength the work started to roll along. The information that they were generating from other genetic markers showed that the gene was towards the middle of the chromosome. It also became clear that more than one gene caused asthma and allergy.

Mice led the next step forward. John Todd in the lab downstairs was studying diabetes, in humans, but also, very successfully, in mice. Beacuse he had showed it was possible to map the genes for a complex disease in animals, I had become interested in doing a similar mouse-breeding experiment for the allergic antibody, immunoglobulin E.

I had made a literature search for all papers written in the last five years on the control of immunoglobulin E in mice. The result was a thick sheaf of pages printed with 'abstracts', as summaries of research publications are called. It was early evening, and everyone had gone home. This was always a good time to think, as the phone did not ring and no-one was asking me difficult questions to which I did not know the answer. I made a cup of coffee and began to skim through the abstracts, noting which ones seemed interesting enough to go back and read in detail. I was reading quite fast and could easily have missed the sixtieth abstract, but for some reason I did not. The abstract described the localization of a gene for the high affinity receptor for immunoglobulin E to mouse chromosome 7.

The high affinity receptor is a fascinating molecule. It is the trigger that fires off the allergic response. Without it there would be no allergy. Julian and I knew about the receptor, but did not think it was relevant to our work because it was located on human chromosome 1, not on chromosome 11. What we had not thought through was that the high affinity receptor was made of three different proteins, called the alpha, beta and gamma chains. The genes for the alpha and gamma chains were on chromosome 1, but the chromosomal localization of the human beta chain was unknown.

What the abstract I was reading said was that the beta chain in mice was close to a gene called CD20, and this is what made me sit up. CD20 in humans was on chromosome 11, very close to the atopy gene. Because mouse and human chromosomes are only rearrangements of the same basic structure, the beta chain, quite possibly the trigger for the whole allergic process, was almost certainly bang in the middle of the map for our asthma gene. I did not jump up and down with excitement. I just thought how extraordinary it was that this gene was sitting on chromosome 11 in just the right place for asthma. I read the abstract again carefully, in case I had missed something. The abstract still said the same thing. It all had the feeling of truth about it: the beta chain must be the asthma gene; it did exist and it was beautiful.

The human gene for the beta chain had not been cloned, or if it had it was still a secret. The people who were likely to clone it were Henry Metzger and Jean-Pierre Kinet, who worked in the National Institutes of Health in the United States. By the calibre of the work they had done on the high affinity receptor they were obviously rather smart characters. Shirakawa arrived from Japan and immediately set to work like a Japanese. He obtained sufficient human sequence for Andy Sandford to prove that the human beta chain was on chromosome 11.

The full human sequence was then published by Kinet. To prove that the beta chain was the atopy gene we had to show that it was different in people who were normal and people who had atopy. Shirakawa worked harder than ever, often eighteen hours every day, spending money like water. We didn't care. We had to know if we had found the gene.

The work went forward very slowly. We had moved from a position of not knowing anything about genetic linkage to not knowing anything about large scale sequencing. We also had no experience of cell biology, so we could not really understand how a cell surface receptor containing variants in the protein made by the beta gene might be expected to cause asthma.

About this time I went to a remarkable lecture. The lecture helped me understand more about cell surface receptors, but the reasons for my always remembering it are more to do with the loftier aspirations of humanity and of science. The lecture was given by Professor Alan Williams, an immunologist from the William Dunn School of Pathology in Oxford.

The lecture theatre was packed. The speaker was a short man, strongly built, with sandy hair and round glasses. He was older than most of the audience, but could not have been more than 45: vigorous to look at, in the robust prime of life. McMichael had been to some pains to find him water to put on the lectern, in a coffee mug. The last of the audience settled, those who could not find seats arranging themselves on the stairs. The speaker began by apologizing for the title of his talk, which was too pretentious; he just wanted to give us an understanding of his research, now totalling twenty years of work. His voice was plainly Australian, honest and straightforward. A slight catch to it explained the water, the remains of a cold perhaps. He told us of his work, the systematic search to understand the molecules on the surface of cells. Through these molecules, he explained, cells communicate in the myriad interactions demanded by the miracle of our existence. From the discovery of the first simple proteins he led us through molecules of increasing complexity of both design and function. He showed us that, as he spoke, he was tracing the paths of evolution, the paths taken by nature as she experimented and built organisms of increasing sophistication and beauty.

He smiled often during the talk, usually at his own past mistakes and misapprehensions. Not only an exceptional scientist, he was also an exceptional man. It was not an ostentatiously polished dissertation, but organized and compact like the speaker himself. Shining out from the composed figure at the lectern was an extraordinary power of insight and understanding. Towards the end, he allowed himself to conjecture at the many remaining mysteries, as speakers usually do at their peril. But each novel idea, drolly presented, conjured the music of the spheres, the fundamental truths of nature. It became clear we were in the presence of greatness, a man in total sympathy with the forces of biology he had studied for so long.

After an hour, the talk finished to loud applause, almost too loud perhaps. He had not touched the water. The audience were permitted questions. He answered in depth, his replies tumbling out so fast that he was almost short of breath. What was the hurry? He was only 45, so there would be other lectures and other audiences. It was because of his passion for his subject: he had so much to say and time was running out; the clock had already passed the appointed hour. One last question, one last answer. The press of ideas was bubbling forth; so much that was unknown, fascinating, strange, beautiful: another twenty years work at least. The clock moved on. He finished in a sea of applause. Only now, at the very end, did he give a sad wry smile. The cancer in his chest was beyond cure. There, before those of us privileged to be with him, he was saying goodbye to the science he loved.

Because the beta gene protein was a cell surface receptor, I contacted Alan Wiliams after his lecture, to seek his advice. He was frantically busy, trying to finish a very important book at the same time as receiving chemotherapy and being assessed for a possible heart–lung transplantation. He said he would be delighted to see me, but the only time that he had would be when he was in hospital for therapy.

I visited him a week later, finding him sitting in bed, working through a pile of papers, ignoring the wires and tubes attached to his body to relieve his pain. He was genuinely interested in the research, reading through the sequence of the mouse beta gene, translating its code into the amino acid sequence as a Greek scholar would translate an ancient text. He showed me how similar the beta protein was to the CD20 protein which arose from the same part of chromosome 11, explaining that they were part of the same family. The shared sequence between members of gene families was always more obvious in the amino acids than in the DNA because evolution tolerated mutations in nucleotide sequence better than changes in amino acid sequence. He told me how the immuno-globulin receptor would be assembled by the cell, and how its function might be investigated. He introduced me, for the first time, to the idea of 'polymorphisms' in the beta gene; that is to say, that we were probably searching for variants of normal rather than mutations which would cause the gene not to function.

In the hour we were together I learned an extraordinary amount. When we had finished, I told him that I would let him know how we got on. He just smiled. 'Good luck with the polymorphisms!', he smiled again as I left, and turned back to his work. Three weeks later he was dead.

Now, with the end potentially in sight, we were overtaken by catastrophe. In March I flew to Canada to discuss a possible collaboration with geneticists in Toronto. The flight home was awful. Both men in the seats next to me had paid about half the fare I had. A small boy a short distance in front had shrieked all night, perhaps because he knew that his father, his only company on the flight, had no idea of how to comfort him. When I had fallen asleep I had been woken by the distinctive sound of someone vomiting in the next row. The flight was remarkable for its lack of turbulence, so it must have been something she had eaten, or drunk. At journey's end the boy left the plane blissfully asleep in his father's arms, and I went on to work.

When I arrived at the hospital I was told that there had been press interest in our research. This had been the result of general curiosity about the Wellcome Trust, who were selling a large part of their shares in Wellcome plc, the pharmaceutical company. The genetics of asthma was to be background to a piece about the Trust itself, buried somewhere on the back pages. In the evening the journalist from *The Sunday Times* rang and asked a few innocuous questions about allergy and the environment. I knew she had spoken at length to Julian, and that she was going to read him the article over the telephone before publication. I went gladly to bed.

The following morning and afternoon my thoughts were on international rugby, as I watched England trample Wales on their way to the grand slam. At ten in the evening I was rung by a journalist from somewhere foreign, who wondered if I was one of the doctors who had cured asthma. I assumed that they had made a mistake. The phone rang continuously thereafter, the only call sticking in my memory being from a distraught Julian, who had spent most of the evening talking to *The Sunday Times*, attempting desperately to get them to change their story. This, he could tell me, went along the lines of 'Asthma gene found, asthma eradicated'. Either the journalist had not read all her copy to Julian, or her story had been added to after she had spoken to him.

The combination of allergies and genes meant that the story was picked up in newsrooms all over the world. The simplest thing to have done would have been to say that *The Sunday Times* had made it all up. The problem was that a potential asthma gene, that for the beta chain, did exist, and that the report contained sufficient elements of truth to make it difficult to deny out of hand.

The responsibility for the whole episode, and for the distress that it eventually caused the parents of so many asthmatics, is nevertheless ours. In general, doctors or scientists who speak to the media need to understand that the media have their own agenda. It is one thing to be enthusiastic about one's research to colleagues, who know enough to take it with a pinch of salt; it is another to enthuse to a reporter, who may either have insufficient knowledge to be cynical, or who may be easily carried away by the eagerness of others.

On the Monday I was rung by Marcus Pembury, the Professor of Genetics at the Institute of Child Health in London. He was part of a group, led by a dermatologist called John Harper, who had set up a study into the genetics of eczema. Eczema was supposed to be part of atopy, so they may have been studying the same disease as we were in Oxford. They had set up their study carefully, and their results should have been reliable. Pembury told me that their lod score with the marker on chromosome 11 was minus 10, which was as negative a result as you could ask for. Either atopic eczema was a different disease to atopic asthma, at least some of the time, or we had made some serious misjudgments of our data. This made a suitably dramatic contrast to the fulsome praise of our non-result echoing around the newsrooms of the world.

By the end of the week, Richard Smith, editor of the *BMJ*, had criticized the 'unholy alliance' between scientists and the media. His point was entirely valid. Soon after that, a television programme called *Hard News* admonished *The Sunday Times* in a fairly mild way for their coverage of the asthma story. They also attacked the paper at somewhat greater length for their sensational claims about an anti-viral drug, called Acyclovir, used for treating secondary infections in people with AIDS.

A month later two letters arrived. One was from the *Lancet*, who had accepted our paper on the maternal inheritance of asthma; in normal circumstances this would have been wonderful news. The second letter was from Stephen Holgate, a leading British experimental allergist. He was interested in genetics, and had begun his own studies in Southampton, with Newton Morton. He was writing as the co-editor of *Clinical and Experimental Allergy*, a very distinguished scientific journal. The envelope contained four papers from different groups in different parts of the world, all of whom had looked at families with atopy and been unable to replicate our result. There was also a pugnacious editorial from Newton Morton, who had reviewed the studies using phrases such as 'logical disjunctions' and 'Oxfordshire genes'. We were in trouble. It was all made much worse because, a short time before, the whole world had heard the names of Hopkin and Cookson associated with a claim to have found 'the asthma gene', and to have cured asthma.

Later that week, at BMA House in London, Richard Smith told me that *The Sunday Times* was taking the *Hard News* programme to the Broadcasting Complaints Commission. This put an altogether new and more sinister twist to the story. *The Sunday Times* wanted *Hard News* to know who was boss. This was genuinely nasty. The *Hard News* journalists had made a responsible job of restoring the balance to the *Sunday Times* article. If it came to a fight, and it would be morally indefensible to run from a fight, then we could again be in the gunsights of the media, this time as fraudulent scientists. Before setting off for Paddington Station, I sat among the lovely plane trees and the unlovely tramps in Tavistock Square and tried to make sense of the chaos.

By now all the elements for a major crisis were in place. The first concern was scientific. Had we got it all wrong? Was all our evidence coming to nothing? If we had got it wrong, then we had been extraordinarily stupid. Nevertheless, this was a possibility. Had there been fraud? *Nature* was always full of the most horrible stories about scientific fraud. Fraud and litigation with *The Sunday Times*. Oh my dear God, please no. We had replicated the linkage twice. The third piece of evidence, Shirakawa's study, had been even more stringent, but it had not yet been published. Was it flawed in some way we did not know about? And then there was the maternal inheritance. How could we find in our data that the atopy gene on chromosome 11 was maternally inherited, and then discover that several other groups, without any chromosome 11 data at all, had showed an exactly analogous result? It would be an extraordinary coincidence. And the other markers on chromosome 11, which were apparently closer to the gene than the first probe, were they really closer, or had we pushed and polished the data so much that they just appeared to be closer? I did not think so, but I did not know. And finally there was the beta chain, the wild card in the pack.

I went back to Oxford, and again went through our data, breaking it up into groups depending on how and why it had been collected. The results were the same no matter how the data was divided and who had collected it: there was linkage to chromosome 11 from the mother's side.

I wrote to each group who had reached negative conclusions, asking if they would let me see their data. Three of the four allowed us to see their raw results. This is in the best scientific tradition. Two studies were tiny, a third contained a significant mistake. They did not after all contradict our findings. The dermatologists also let us see their eczema data. Their study was big, but, just as Pam had found with our asthma families, nearly everyone in their families was affected. This had reduced the power of their results just as it had with Pam's: a maternal pattern was also detectable, though only faintly. Despite their flaws, these 'negative' studies were widely quoted, mostly by people who could not understand the complexity of the genetics. The general sentiment became that we were wrong about the chromosome 11 linkage. I found it difficult to open my mail, in case it contained more bad news.

Shirakawa worked day and night for over a year. He had sequenced the gene from a dozen people, and had not found a single mutation. He was plainly exhausted. The pressure continued to mount. The Wellcome Trust grant which paid my salary became due. Obviously renewal was not going to be straightforward. After four years of constant scepticism and disbelief we were all feeling very tired. Kafka would have understood, 'Someone must have been telling lies about Joseph K. ....' I began to fathom why innocent people confess to crimes they did not commit when pulled off the streets and beaten in police cells.

Shirakawa had begun at the beginning of the gene and, under Julian's supervision, was working his way through to the end. The gene was made up of seven parts. At the beginning of the sixth he found something. Three bases were different from the sequence published by Kinet. He and Julian translated the code, showing that the mutations resulted in changes to the amino acids that made the beta chain. They were the right kind of changes to alter the function of the gene. Even at this stage we could not yet celebrate. Nine months' work followed, testing for the mutations in families. To be absolutely certain of the results we did all the experiments double blind: the people testing for the mutation did not know from whom the DNA came.

One day in midsummer, the sky for once was blue. We prepared the results and fed them to the computer. In a set of random samples, the result was positive: the level of immunoglobulin E in the blood was related to the presence of Taro's mutations. In families with asthma 15% had the mutation. The statistics showed this could not be due to chance. We could estimate that other mutations could bring the figure up to 50%. There were other asthma genes, but we had caught the first.

CHAPTER 11

CANCER

Cancer is the most feared of illnesses in our society. This fear is to some extent unjustified; many cancers are chronic illness that can respond completely, or partially, to treatment. Nevertheless, cancer will affect many of us or our relatives: 20% of people in our society die of cancer. Half these will succumb to cancer of the lung, breast, or lower bowel.

Cancer is often presented to the public as a single disease. Commonly this distortion arises in the context of a cure for cancer, the implication being that a magic bullet will provide a single cure for all cancers. Quite frequently the presenters of such simplifications can profit from the misapprehension, perhaps to hawk an unlikely cure to those who are desperate for hope. It is therefore important to realize that cancer is not one disease, but many.

One characteristic is, however, shared by many forms of cancer: they are further examples of illnesses in which the external environment conspires with an internal genetic susceptibility. Both environment and genes are necessary to produce disease. Cigarette smoke is the single most important environmental carcinogen (cancer-causing agent). People who do not smoke are one-fortieth as likely to develop lung cancer as those who do. Nevertheless, not all smokers get lung cancer. This is presumably because their genes have protected them against the relentless chemical assault of cigarette smoke. Besides chemicals, cancer can be caused by other things in the environment, such as radiation and viruses. As we shall see, one of these environmental factors, virus infection, has helped unravel the genetics of cancer in a completely unexpected way.

Cancer is genetic in several senses. Like other genetic illnesses, some kinds of cancer run in families – but most cases of cancer are 'sporadic', that is they occur without a family history. However, these cancers are also genetic in another sense, in that they are due to acquired disorders in the genes that regulate cell division and growth.

Although all forms of cancer are characterized by uncontrolled growth of cells, cell division is an entirely normal phenomenon. The old man and the infant in its mother's arms have both grown from a single cell, the fertilized egg. Billions of divisions and differentiations and migrations later that single cell has become a recognizable human being. In all of this staggering intricacy, each division of every cell has been under strict control: a lung or a liver grows to a certain size, and then no larger.

In adults the genetic program for our growth has led the disposition and number of our cells to a stable state; but the stability of our external appearances belies the enormous amount of cell growth that continues in our bodies. Even in the elderly, the linings of the lung and intestine are replaced every fortnight; our red blood cells last only thirty days, and in the same period the turnover of white blood cells may be in the billions. Our lungs and intestine, however, retain their appearances, and in the absence of infections, the numbers of white and red cells remain within narrow limits. When we cut ourselves, cells on either side of the cut immediately begin dividing. When the wound has healed, division stops.

This regulation of growth is achieved throught an elaborate system of signals between cells. From what is known of these systems, and what is now known can be only a fraction of the whole, the signals are often in the form of soluble proteins.

One example of these proteins are the 'growth factors' which are released from one cell to induce a response in a neighbouring cell of the same or different type. Cells select the factors to which they respond by means of a 'receptor', another protein, or complex of proteins, which sits on the cellular surface membrane. The growth factor fits the receptor like a key in a lock. Cells with the wrong lock do not respond to the growth signal. The activated receptor sets off a chain of events in the cell that induces or stalls cell division or growth, or induces other locks and keys to open other doors to make the cell do other things.

The cells of the immune system also communicate with each other via signal proteins, called 'cytokines', which affect cells in their neighbourhood; though the cells do not have to be touching each other, as is the case with growth factors. Similar signals also act over large distances, when the signal molecules are known as hormones. Science has known about hormones since the first half of this century, but the discovery of cytokines and growth factors has occurred primarily in the last decade. The rate of discovery of new growth factors and cytokines is feverish, because it is driven by market forces. A patent on a cytokine that greatly affects sick patients may be worth a billion dollars a year.

It is not only biotechnology companies that have found the cytokine network profitable. Cytokine receptors on the surface of white blood cells are commonly hijacked by viruses in order to gain entry to the cell. One particularly ingenious virus, the Epstein-Barr virus, employs a variant of the basic hijack technique. In Western society the virus causes glandular fever, the so-called kissing bug, and by the time of adolescence half the population has been infected. A second wave of infection follows, as teenagers start to kiss each other in preference to their parents. In Africa, where nutrition and infection are everyday matters of life and death, the Epstein-Barr virus causes a nasty cancer, called Burkitt's lymphoma. Infection with the Epstein-Barr virus, and indeed with many other viruses, is life-long. Between 15% and 20% of us are 'heavy shedders' with infectious Epstein-Barr virus detectable in our saliva; the heavy shedders ensure that the virus remains in circulation in the population.

The virus's long relationship with humanity has given it a human gene. This gene mimics an important human cytokine, called interleukin 10, and infection with the virus results in an outpouring of the viral interleukin 10. The immune response to the infection is blunted, and the cell infected by the virus is encouraged to divide faster – two actions ideally suited to help the virus succeed in living on happily in its host.

It is highly unlikely that the Epstein-Barr virus has evolved its own interleukin 10 gene; it is much more probable that, at some time in the past, it 'stole' the gene from a human cell. There are half a dozen other examples of genes for cytokines or other immune modulators which have been purloined by viruses. But, in general, these stolen genes have relatively benign actions. However, there are other viruses whose actions may be much more sinister.

A cancer occurs when a cell somewhere in the body escapes its normal mechanisms of control. This subversive cell divides steadily until it has formed a mass, a growth, that is big enough to be noticed, or to make itself noticed. A growth of this kind is called a tumour. Some tumours grow very slowly, and, apart from the fact that they continue to grow, remain under a semblance of normal regulation. These tumours are called benign, because if they are removed they will not recur. Other tumours grow much more rapidly, even outgrowing their blood supply, and spread or metastasize through the body. Such growths are called malignant. The term 'cancer', in lay use, covers any malignant growth. When malignant growth takes place in white blood cells the disease is called leukaemia, when it occurs in lymph glands it is known as lymphoma, and when found in connective tissues such as muscle or bone, it is called sarcoma.

Because of the link between cancer and viruses, the discovery of the genes for cancer followed a totally different path to other gene hunts. The story of cancer and viruses, some of which I will summarize here, is wonderfully related by George Klein, in 'The Tale of the Great Cuckoo Egg' from his book *Ateisten och den Heliga Staden* (The Atheist and the Holy City).

The scientific adventure of viruses and cancer began with an American called Peyton Rous. In 1911 Rous showed that he could transmit sarcoma from an affected chicken to an unaffected chicken. He broke up the sarcoma in a blender mechanism, and filtered it. The liquid which passed through the filter contained no cells, because the pores of the filter were too fine to permit cells or bacteria to pass. Rous then gave this 'cell-free filtrate' of the tumour to healthy chickens. A few of these chickens also developed sarcomas. Viruses were undiscovered, and so Rous could not explain his findings; he could only say that something in his filtrate could cause chicken sarcoma. Other scientists found that other tumours were not transmissible in this way, and the results, although interesting, did not have a direct application. Consequently, Rous' work was all but forgotten for decades.

In the early 1930s scientists at the Jackson Laboratory in the United States began to breed mouse strains which had a high or low incidence of cancer. After producing many generations of mice they succeeded in developing strains that were susceptible to various types of malignancy, and other strains that were highly resistant. These results meant that cancer, at least in part, was under genetic influence, and experiments in cross-breeding the high and low cancer strains showed that it was polygenic, that many genes caused it rather than one or two.

A critical experiment with one of these intercrosses showed that the risk of breast cancer in a particular kind of mouse was only increased if it inherited the cancer genes from the mother's side. In the 1990s, a geneticist, aware of arcane mechanisms such as genomic imprinting, would consider several possible explanations for this maternal effect.

Fortunately, the Jackson Laboratory geneticists knew nothing about genomic imprinting, and started to look for the infectious agent which transmitted the cancer from mother to child. A researcher called John Bittner isolated the responsible substance in the milk from affected mice. He called the agent 'milk factor', because the prevailing belief was that viruses did not cause cancer, and he realized that if he had claimed to have found a cancer-causing virus he would have come under severe attack. Today, however, the milk factor is known as mouse mammary tumour virus, or Bittner virus. The milk factor was not enough to cause the cancer on its own, but it quadrupled the risk of cancer in susceptible mice. Bittner's virus then followed the fate of Rous' sarcoma factor and was forgotten. Again, the results were too far ahead of the general understanding that would have highlighted their significance.

George Klein has described the enormous increase in knowledge that followed, as the genetics of cancer in mice were investigated over the next twenty years. These studies showed that cancers were due to a variety of genes, acting in different tissues at different times. Although they did not identify the genes themselves, they prepared the way for the surprise that the viruses were to deliver, the surprise that Klein has called 'the cuckoo's egg'.

Twenty years after Bittner's discovery of his milk factor, another researcher, called Ludwick Gross, found he could transmit mouse leukaemia with a cell-free filtrate. Gross used a particular strain of leukaemia-susceptible mouse and, because he had decided it was important to escape the mouse's immune system, he injected the filtrate when the recipient mice were less than 24 hours old, before the immune system was mature enough to react against viruses. The transfer of cancer in a cell-free filtrate from one animal to another echoed the work of Rous some forty years before. This time, however, no-one could repeat Gross's experiments. Gross was an émigré to the United States. He did not speak English well, and he worked as a technician because his qualifications were not recognized. Regrettably, the consequence was that Gross became the subject of general opprobrium: his experiments could not be replicated, and therefore the only conclusion was that he had cheated by making up his results.

Finally, after five years of misery for Gross, Jacob Furth, the scientist who had bred the leukaemic mouse, attempted to repeat Gross's experiment. In contrast to those who had gone before him, he carried out the protocol exactly as Gross himself had done, injecting new-born mice who were less than 24 hours old. He achieved transfer of the leukaemia just as Gross had said.

One can easily imagine how unhappy Gross must have been during the previous five years. However, now that the reputable Furth had repeated the experiment, the results were impossible to ignore. Within a year many other viruses causing cancers in different animals had been isolated. The concept was soon formed of an 'oncogene', a cancer gene, carried by the virus. The results were so exciting that the results of the previous twenty years' breeding experiments were forgotten. The pendulum of opinion swung to a position where all cancer was due to viruses.

In the 1960s it became possible to examine the virus genes directly. The oncoviruses were retroviruses, that is their genetic code is made of RNA, in a single strand. Retroviruses make an enzyme called reverse transcriptase, which can insert the virus genes into the DNA of the cell which they have infected. The virus genes are then part of the DNA of every descendant of the original infected cell. Retroviruses are very simple in structure, and usually have only three genes. All of their other needs they borrow from the cell they have infected. They even wrap themselves in the cell membrane rather than making their own coat.

Sequencing the Rous virus for chicken sarcoma resulted in the discovery of a fourth gene besides the normal three. Was it possible that this fourth gene was an oncogene? This very exciting hypothesis was proved to be correct. The fourth gene caused the chicken cancer and was christened with the abbreviated name of *src* (for Rous sarcoma gene), which is pronounced 'sarc'. The first cancer gene had been discovered. As a result Peyton Rous won the Nobel prize, in 1966, at the age of 86. He had waited 55 years for the rewards of his experiments.

Now that the first viral oncogene had been isolated, other oncogenes were quickly found from other inbred animals and other types of tumour. It is completely remarkable that these oncogenes were, in every case, identical or nearly identical to a normal gene, found in the normal tissues that the virus had infected.

George Klein asks a very interesting question about these oncogenes. What are they doing in a virus? In the wild state, they offer no evolutionary advantage to the virus. Inducing cancer does not help the virus to replicate or spread to other hosts and a useless extra gene is an enormous burden for such a simple virus to carry, as it increases its genetic material by a quarter.

Klein then makes a brilliant suggestion: that the extra oncogene is in each instance the result of an accident. Virus genes move in and out of their host cell's DNA; this movement of genes is an essential part of the 'life cycle' of the virus. Indeed, retroviruses, like the cancer viruses, may have begun their existence entirely in the host genome, as the transposons described in 'DNA' (see page 15 ff.), and only escaped the cell late in their evolution. As viruses move their genes out of the host DNA (to form a complete virus that can infect other cells), a mistake will occasionally occur, so that a piece of the host genetic sequence is caught up in the viral genome. Very occasionally an entire gene is caught up.

The force of evolution means that a cancer gene will only be retained in a virus if causing cancer gives some advantage to the virus. In 'normal' circumstances the cost of carrying the extra useless gene means that it will quickly be lost from the virus. Obviously, causing cancer does not help a wild virus survive in normal mice. However, in the Jackson Laboratory in the 1930s, mice were being deliberately bred to develop cancer. Thus, paradoxically, the mice that were susceptible to cancers stood the best chance of passing their genes on to other generations. The laboratory breeding was imposing tremendous evolutionary pressure to bring out cancer genes from amongst all the normal genes in the mice. The process was successful. High-cancer strains of mice contained many cancer genes in their chromosomes.

To recognize this effect on the mouse genes was to do no more than to say that the purpose of the experiment had been achieved. But what of the viruses? The evolutionary pressure provided by selective breeding in the laboratory had had an unexpected side-effect. As well as selecting in favour of mouse genes for cancer, the breeding programme had forced the evolution of viruses that both infected the mice and caused them to acquire cancer.

In all, about twenty viral oncogenes have been isolated. Because the virus genome is so simple, and because RNA from the viruses could be easily purified in large volumes, the sequencing of the oncogenes was achieved twenty years earlier than might have been possible with positional cloning and other modern techniques. The evolution of viral oncogenes is George Klein's 'cuckoo's egg', the surprise in the viral nest.

Sequencing the viral oncogenes made it possible to clone their normal animal counterparts. It was discovered that oncogenes were usually parts of the pathways that controlled cell growth. In a normal cell the genes were turned off. In cells infected by virus, the genes operated continuously, and as a result the cells divided indefinitely. Sometimes the oncogene was shown to be a mutated version of the normal gene, so that a normal action to suppress cell growth was lost, and malignant cell division took place by default.

More orthodox genetics has also helped unravel the mysteries of cancer. Several malignancies, particularly leukaemias, have a distinctive pattern of fragmented chromosomes. Chronic myeloid leukaemia (CML) is a cancer of a type of white cell, which grumbles along for years before entering a final stage of heightened malignancy. In the white cells of people with this leukaemia there is often an abnormal chromosome 22. This abnormal fragment is called the Philadelphia chromosome after the place of its discovery, and is usually joined to a small abnormal piece of chromosome 9.

It was years after the discovery of the Philadelphia chromosome that an oncogene, c-abl, was found on chromosome 9. The break of chromosome 9 in chronic myeloid leukaemia decapitated c-abl. The decapitated c-abl was joined to the head of a gene called bcr from chromosome 22. Together the two genes make an abnormal protein that forces the cell to divide continuously. Uncontrolled division is the cause of the leukaemia.

In another type of leukaemia, a break in an oncogene on chromosome 11 allows it to join a gene that normally makes antibodies. Many white blood cells are antibody factories, pouring out enormous amounts of antibody proteins to combat infection. In the leukaemia cell, the chromosome 11 oncogene is turned on as if it were an antibody gene. The result is a continuous discharge of oncogene, continuous cell division, and cancer.

These breaks and rejoinings of chromosomes, called 'translocations', evidently do not occur entirely by chance, as it has been found in Cambridge that the DNA sequence in the two breakpoint regions is often similar. The normal mechanisms of repair then splice the chromosome fragment into the wrong place, with the resulting catastrophic consequences.

Some types of cancers run in families, and so the search for new oncogenes has focused on the study of such families. Familial cancer, however, accounts for only a few percent of the total number of cancers (although many people feel that cancer runs in their family, this is usually just because cancer is so common: after all, one person in five dies of some sort of malignant growth) and most cancers are 'sporadic', occurring without an obvious hereditary predisposition.

A key observation, from families with genuine hereditary cancer, was that individuals who carried a cancer-causing gene did not always develop cancer. Often, the appearance of cancer was delayed for forty or fifty years. Clearly, a single mutation in a single gene was not sufficient on its own for a cancer to begin. What then was happening?

Familial adenomatous polyposis coli (APC) is a hereditary disease of the large bowel. It is inherited as a dominant single gene disorder, so half the children of an affected person will be diseased. Sufferers develop many polyps, small growths, on the lining of their intestines. Over the years the number of polyps increases, but they remain benign, growing to a certain size and then stopping.

Some individuals with the condition lead almost normal lives. However, in many people something else happens: one polyp develops into a cancer.

Because APC is a dominant condition, it was very amenable to the reverse genetic approach. As with other conditions, many groups in many centres competed and cooperated in the hunt for the gene. In 1992, teams led by Yusuke Nakamura at the Cancer Institute of Tokyo University, and Bert Vogelstein at John's Hopkins in Baltimore, found the APC gene. Their discovery has explained much about the nature of cancer.

The cells that line the intestine are incessantly growing and dividing, as the intestinal surface is constantly replaced. This exuberant production of new cells is, however, confined to 'crypts', the bottom of deep folds in the lining of the bowel.

Mutations in the APC gene prevent it from working properly. When this happens, the proliferating cells grow up out of the crypts onto the surface of the intestine. This increase in cell division is in itself harmless, except that with each cell division there is a very small chance of a genetic mistake in the other genes in the cell.

Occasionally, amongst the billions of dividing cells, there is a mutation in another oncogene, this time called k-*ras*. Mutations in k-*ras* cause the gene to become abnormally active and the nature of the affected cell and its descendants to change slightly, with the polyp containing the cells becoming larger and more pronounced. Cell division is more active, and there is thus even more opportunity for genetic mistakes.

Later on another gene, called *DCC*, is lost, and cell growth accelerates again. At this stage the growth is still benign, although the polyp is no longer organized as neatly as in normal tissue and the capacity for genetic error has increased yet again. Finally a critical gene, called *p53*, is lost.

This last gene, *P53*, was recovered from an oncovirus in 1979. Like other oncogenes, it seemed to encourage cell growth and it took another ten years to discover that the viral form of *p53* was a mutant. When the normal *p53* was found, it had completely the opposite action to the mutant – it acted as a powerful brake on cell division. *P53* is now the queen of the oncogenes. It is damaged in many types of malignancy: two-thirds of lower bowel cancers, half of lung cancers, and a third of breast cancers all have mutations in *p53*. If *p53* is abnormal, these tumours are likely to be more agressive, and the outlook for the patient is worse than if *p53* is undamaged.

In the absence of normal *p53*, the growth of cells in the intestinal polyp is unrestrained. The cells begin to spread through normal surrounding tissues deep into the wall of the bowel. A cancer has begun. Later still, other genes may fail, allowing the cancerous cells to break away and spread through the body.

Thus, for a cancer to form and spread, between five and seven genetic 'hits' must have occurred. When other colon cancers, which had appeared in people who did not have hereditory APC, were examined genetically, the APC gene was still abnormal. The mutation had not been inherited; it had happened by chance in a dividing cell, but the consequences were the same.

Other cancers show the same pattern of 'multiple hits'. Carcinoma of the cervix, the neck of the womb, is one of the few human cancers to be definitely caused by a virus. The virus, human papilloma or wart virus, is transmitted sexually. As a result, sexual activity with multiple partners increases the risk of this cancer. The papilloma virus carries genes for two particular proteins, E6 and E7, which inactivate the *p53* gene and another tumour suppressor gene called *RB*. Switching off these genes is advantageous to the virus because it allows it to replicate more freely within infected cells. A side-effect is that the infected cells have lost two protective growth repressor genes, the equivalent of two hits. Chance mutations in other important genes can then cause cancerous growth.

We find a further level of complexity in the causation of cancer when we look at the most common cancer of women: breast cancer, which one woman in ten will develop by the age of eighty. The incidence of early forms of breast cancer has been increasing steadily for fifty years, although the incidence of the more malignant variants and the rate of women dying from the disease has remained steady. The cause of this increase is unknown, although early diagnosis may play some part.

Breast cancer is unusual, but not unique, in that the tissue of the normal breast responds to hormones. (Other tumours show patterns of hormonal responsiveness, most particularly cancer of the prostate.)

The relatively modest hormonal fluxes during a normal menstrual cycle influence the breast, as do the much larger hormonal changes of pregnancy. It has been suggested that rhythmic hormonal changes during the female ovulatory cycle predispose to breast cancer. This is because the increase has come about at the same time as an earlier and earlier onset of puberty.

Hormones are also implicated in breast cancer by evidence from women who have had their ovaries removed before the normal menopause: they carry a lower risk of breast cancer than other women. On the other hand, pregnancy protects against breast cancer. Women who have never had children carry a higher risk of breast cancer than those who have given birth. This could mean that pregnancy takes the tissues of the breast through their full pre-ordained path of normal growth, lactation, and then involution to a resting state.

The risk of breast cancer associated with not bearing children is not terribly high, only two or three times normal. The risk associated with an early puberty is less than twice normal. These associations, and others like them, should not therefore cause women to rush to fall pregnant, or to starve their children to delay the onset of puberty. Because the potential for growth is intrinsic to the normal function of the breast, which is lactation, the tissue of the breast carries an especial risk of malignant change. The inbuilt responsiveness to hormonal stimuli may be thought of as the equivalent of a 'hit' in a cancer suppressor gene, in that there is one less check to abnormal growth available.

Responsiveness of the breast tissue to hormonal influence is often retained in malignant breast cancer, which allowed the introduction of a revolutionary treatment for breast cancer into clinical use around 1980. This drug, called Tamoxifen, is an anti-oestrogen with very few serious side-effects. It blocks the oestrogen hormone receptor on the breast cells, and removes one of the stimuli for abnormal growth. Although it will not repair the 'broken' oncogenes that led the cancer cells to divide uncontrollably, it may restore the balance in favour of non-growth sufficiently for cells to involute and the tumour to shrink.

Not all women with breast cancers respond to Tamoxifen. Those who do not are often lacking a normal oestrogen receptor on their cells. In these tumours a mutation may have occurred in the receptor gene itself, or in a gene that controls its appearance on the cell.

Although various hormonal states somewhat influence the risk of breast cancer, one factor increases the risk of the disease a hundred-fold: a family history of the illness. This means that a particular gene may predispose to the disease. Five percent of women with breast cancer have the familial variant, so that half a percent of women carry a breast cancer gene. Of those who carry it, four-fifths will develop cancer. In Britain there may therefore be 250,000 females with the defective gene, and 200,000 of these are likely to develop breast cancer. A gene for breast cancer, of a particularly aggressive type which comes on in early middle age, was localized to chromosome 17 in 1990. The localization was the result of fifteen years of work by Mary-Claire King, a Californian epidemiologist who had turned to genetics. The elements of the search are similar to those of other searches: international collaboration, protracted hard work (linkage was found with the 183rd marker), and an initial general scepticism about the result. This gene has not yet been cloned, but the bulldozer, Yusuke Nakamura, and other intensely competitive groups in England and America are only a short distance from success.

The successive discoveries of the way in which genetic hits contribute to cancer will soon make an enormous difference to cancer care. At the moment, doctors can only guess the future behaviour of a cancer by looking at a piece of the tumour under the microscope. To be able to say that a particular oncogene, or subset of oncogenes, has been mutated will give much more precise information. It will enable us to predict how the tumour will behave, and how it will react to particular treatments. Surgeons will know much more accurately the type and extent of operation they need to perform to remove the cancer completely. With certain combinations of oncogenes they could say that an operation will be ineffective; however, even with these patients, a full knowledge of the mechanisms of cell growth in the individual cancer will mean chemotherapy can be tailored to maximum effect.

The expansion of our knowledge of oncogenes and oncogenesis has opened the way for the demise of a particular theory which I have never liked. For years scientists and doctors who study cancer have been taught the doctrine of 'immune surveillance', which holds that potentially cancerous cells are continuously appearing in the body, and that the immune system in some unknown way keeps watch for these cells and destroys them. The hypothesis had its origins in early experiments when tumours were transplanted between different strains of mice. In the host mice the tumours shrank and disappeared. The conclusion was that cancer could be overcome by the immune system.

There are problems with this argument. Firstly cancers, even highly malignant ones, are made of the same constituents as the normal cells of an individual. It is highly unlikely that the immune system will recognize these cells as foreign and destroy them.

Secondly, the hypothesis of immune surveillance is hopelessly simplistic. It is easily possible to observe the control of cell division in normal life. If you cut a finger you bleed and there is a gap in your integument, your skin. The gap fills with a clot, made of substances from your blood. Over the next few days fibrous cells, new blood vessels, and new skin grow into and over the clot. When the wound is healed, the new growth of these complex elements ceases. A miracle of coordination between perhaps fifty types of cell has taken place to return your finger to its previous state, and, other than in keeping the wound free of infection, the immune system has played no part. So too, I am sure, with cancer. Cancer is due to broken genes.

Proponents of the immune surveillance theory point to the enormous number of malignant cells that make up a cancerous tumour or fill the bone marrow in someone with leukaemia. Following successful chemotherapy, all the abnormal cells disappear forever. The chemotherapy, it is argued, cannot possibly have killed all the abnormal cells, and so the immune system must have had a vital part in removing the malignant survivors. This is not so. First, there are places where malignant cells can hide from chemotherapy. These 'sanctuaries' include the linings of the brain and spinal canal. The immune system permeates the sanctuaries. Nevertheless, if the sanctuaries do not receive special treatment during cancer therapy, cancer and leukaemia recur in them, and spread out again into the body.

The second misconception is that all the cells in a cancer or leukaemia are themselves malignant. In young men the most common malignant disease is a cancer that arises in the testicles. In this illness the cancer cells spread rapidly through the body, appearing as 'cannon balls' which are inches wide in the lung and the brain. Only twenty years ago, the life expectancy of anyone unfortunate enough to contract this cancer was 24 months or less. Now, in the age of chemotherapy, over 90% of those with the illness are completely cured.

When chemotherapy was first applied to the illness a very strange phenomenon was seen. These cancers released a hormone into the bloodstream, called human chorionic gonadotrophin. The presence of the hormone was a very sensitive indicator that the malignancy was active in the body. Chemotherapy could be considered to have had a successful conclusion if the hormone disappeared from the blood. In many young men the hormone became undetectable after therapy, and their well-being returned to a state in which they felt cured. However, although the masses seen on the chest X-ray and felt in the abdomen shrank, they did not disappear.

At first it was thought that the cancer had persisted, and had changed in nature so that it no longer secreted the tell-tale hormone. However, the masses did not increase in size as the previously malignant growths had done. When they were operated on and examined under the microscope they contained only normal cells, although the cells were not in their normal place.

What had happened? The malignant cells maintained some of their original function. They could signal to the cells around them to multiply and to differentiate as if they were in a normal growing tissue. Although the chemotherapy had removed the malignant cells, the organization they had built around themselves remained, just as the buildings of the Incas persisted long after the Incas themselves had perished. It is likely, therefore, that much of the mass of many malignant tumours is not made up entirely of malignant tissue.

There is one more twist to understanding cancer. We have seen that cancer is usually considered to be the result of abnormal growth, but something altogether more subtle and startling is also at work. Developmental biologists study how embryos grow; for twenty years they have known that for normal development many cells in an embryo have to die. Only recently has the idea emerged that this death is programmed; it happens under genetic instructions. The cell will, when told to do so, kill itself.

In the nervous system many more neurones (nerve cells) are made than are needed. Only a limited number of these survive. Those that do so are those sustained by growth factors from other neurones. The reliance on growth factors ensures that the correct connections are made between different types of neurones, or between nerves and muscles. Martin Raff, of University College London, suggests that the redundancy in the number of neurones also introduces a critical element of competition between the cells. Only the fittest neurones survive; those with acquired errors in their constitution will perish. This removes the need for developmental genes to work infallibly: mistakes can be tolerated because cells with mistakes do not survive, and their place is taken by a normal competitor.

Cells which are dying on instruction look different to cells which are dying from other causes. Programmed cell death is called 'apoptosis', a word derived from the Greek, which describes the autumn fall of leaves. Apoptosis happens when cells are deprived of hormones or growth factors.

Research at the Walter and Eliza Hall Institute in Melbourne, and Washington University School of Medicine, linked apoptosis to an oncogene called *bcl-2*. Scientists at these institutions put the *bcl-2* gene into white cells. These cells depended on cytokines to stay alive: if they were not given cytokines, they went into apoptosis and died. The *bcl-2* gene prevented suicide in the white cells, even if the cytokines were withdrawn.

Fascinatingly, *bcl-2* was found to resemble a gene from the roundworm. This worm, *C. elegans*, has only 1090 cells, 131 of which die during development. An important part of the Genome Project is to sequence all the worm's genes. It will be the first organism in which all the genes and their functions will be exhaustively worked out. It was noticed that *bcl-2* was similar to *ced-9*, a gene which prevents apoptosis in *C. elegans'* developing cells. The similarity between these two genes, in living creatures as distant as a human and a roundworm, suggests that programmed cell death is an elemental part of cellular behaviour.

In normal cells it seems that *bcl-2* is turned on by cytokines and growth factors. If the cytokines are not present, *bcl-2* is withdrawn and the cell dies. This ensures that a cell that is in the wrong place will kill itself. Other oncogenes play a part in apoptosis: *p53* seems to act in the opposite way to *bcl-2*, and induces cell death when it is turned on; another oncogene, *myc*, also stimulates apoptosis. Mutations in these genes allow cells to survive away from their normal support of growth factors and cytokines.

Thus the understanding of oncogenesis has gone hand in hand with the process of deciphering the normal growth, from a single fertilized egg to the trillion cells of an adult human body. We have begun to realize that cells speak constantly to each other, sustaining or suppressing one another with a continual traffic of signals and receptors. Of all the complex genetic diseases, cancer is the one to the study of which the new genetics has contributed most. With such a multiplicity of new targets, it is inevitable that profound benefits, in the form of new treatments for cancer, will follow.

CHAPTER 12

SCREENING: PANDORA'S JAR

As we have seen, it is now possible, in principle at least, to find the cause for every illness that owes its provenance in whole or in part to variation in our genetic make-up. The genes that cause the major genetic illnesses have been hunted down and their secrets laid bare. The common diseases of diabetes, cancer, and high blood pressure are only a few years away from yielding to the gene hunters. If we are to believe all that we read, a new and wonderful age of molecular medical cures awaits us. The reality, of course, is that finding a gene, or a mutation in a gene, does not lead immediately to a remedy: indeed, a cure may remain impossible.

Nevertheless, capturing a gene that causes a disease invariably opens up new approaches to treatment. These possibilities, broadly speaking, can be categorized into three groups, which can be ranked, possibly in ascending order of desirability, and certainly, for the moment at least, in descending order of practicality.

First, families at especial risk, or the population at large, may be screened for the abnormal gene, and they and their relatives given appropriate advice. Second, identifying a disease gene can define a target for new drugs. Third, as genetic mutation results in an absent or abnormal protein, disease may be treated by substituting the normal protein or even by replacing the faulty gene.

Although this list of possibilities is highly encouraging at first sight, on closer inspection the issues surrounding each prospective treatment are far from clear.

The first therapeutic expectation follows immediately from the discovery of a mutant gene. This prospect is genetic screening. When Botstein and Bodmer were advocating the first genetic map, their chief hope was that it would lead to better screening.

Screening reveals the risk of a person developing a genetic disease. Sometimes this information is of pressing consequence to the person concerned; at other times it is of more interest to others: those in authority, such as parents, health authorities, or insurance companies. Importantly, the risk cannot always be estimated accurately. With single gene disorders, the risk of disease can often be given as nearly 100% or, conversely, as one in a thousand or less. The probability of developing a complex disease can never be given with the same exactitude.

Screening is carried out quite simply. A sample of blood, or the cells from a simple mouthwash, will give all the DNA that is required for the test. The DNA can then be examined for the presence of the relevant mutation. Mutation screening gives the most accurate results, but it depends on knowing which mutations in which genes are likely to cause disease. Screening can also be performed at the stage of genetic linkage, when the genetic defect has been shown to lie within a particular region of a particular chromosome, before the mutant gene is itself found. For screening at this stage it is necessary to examine the parents and other children in the family who are affected and the results can quite often be inconclusive.

It was first assumed that individuals at risk of genetic disease would be overwhelmingly in favour of screening. This has not at all proved to be the case. What geneticists had not realized was that the efficacy of screening depends enormously on whether it leads to a useful treatment.

Huntington's disease epitomizes this problem. The illness usually comes on only after the age of forty, and ends ten years later with complete inanition. By the time they are forty most individuals have fathered or mothered their children, and have committed their genes to the next generation. There is no cure for Huntington's disease. Even the discovery of the gene has not yet delivered any means of relieving an inexorable slide to madness and despair. Assume you are an adult of thirty, and that one of your beloved parents has died of a horrible illness. There is a 50% chance that fate has decreed that you too will develop this relentless and insufferable disease in ten or fifteen years. Do you want to know now, while you are still hale and hearty, that this is to be your destiny? Will you be tested? It is not at all clear that the answer is yes.

Once in this position, other questions tumble out. If you have children and they carry the Huntington's disease gene, do you have the right to tell them what awaits them? Can your family make you take the test if you do not want to? Very importantly, who will help you or your children if the test is positive?

Even if you are convinced of its merit, what happens if you take the test, but change your mind before the results are available? Do you then, against your will, still have to find out? How often are the results wrong or inconclusive?

Balanced against these negative considerations is one undeniably positive benefit of screening for an untreatable condition. If it is shown that you do not carry the mutation, then you and your children will be freed from the unnecessary fear that would otherwise burden you all your lives.

There is no simple right or wrong to genetic screening in these circumstances. An international body, the World Federation of Neurology, has drawn up guidelines for screening for Huntington's disease. These protect the individual as much as is possible from the bad effects of screening. They declare that screening must be voluntary, and not at the instigation of a third party such as a potential spouse or an insurance company. It should only be carried out in centres expert in genetic counselling. Subjects should give informed consent, after all the proper information has been supplied to them in a written and a spoken form. Prenatal screening is permissible, with the mother's wishes overriding those of the father. The subjects should select a 'partner', someone they can trust, such as a spouse or a social worker, to accompany them through the testing process. It was recognized early that individuals who are themselves at risk do not make good partners. The results of testing must be kept absolutely confidential. All possibilities must be discussed with the subject before the test takes place, and he or she should have the right to refuse to know the result at any stage.

These guidelines are eminently sensible, and protect the patient in the best traditions of medical practice. If the result of a screening for Huntington's disease is positive, then the mechanisms for communication and support of the sufferer are already in place. Genetic screening carried out in the spirit of the guidelines becomes just another part of the doctor's armamentarium to diagnose and treat disease. It should be remembered that doctors regularly have to deal with bad news. Even when there is no cure for an illness, such as the advanced stages of cancer or heart failure or a hundred other diseases, the exchange of the truth between doctor and patient is generally far more beneficial than the alternative.

Other equally severe diseases may now be detected by screening. Women who carry a breast cancer gene have an 85% chance of eventually contracting the illness, although, like Huntington's disease, it may not manifest itself before the age of forty. It can be argued, as for Huntington's, that it may be better not to know one's fate. However, unlike Huntington's disease in the present stage of knowledge, breast cancer is preventable. Accurate knowledge of risk means informed decisions can be made about treatment. A choice might be made, for example in the case of a patient who has a low or medium risk of cancer, to watch and wait with regular examinations and mammograms. If the risk warrants, anti-oestrogen or other hormonal agents may be used; or even a preventive mastectomy may be felt necessary. The same applies to other hereditary cancers. Armed with the sure knowledge of a predisposition, it will be possible for the patient and her physician to be much clearer about the steps that should be taken to prevent the disease.

It is also wrong to assume that ignorance is bliss: not knowing that you have a genetic predisposition to cancer will not prevent the cancer from appearing. It is better to be prepared. To come from a family with hereditary cancers means that you will be very likely to have observed the effects of the disease at first hand, and in these circumstances it will be impossible to escape from the fear that you may be affected. If you can be cleared as a carrier of the gene then your life may be considerably easier.

Children have a special place in the screening process. It is generally agreed that children should not be screened for a condition that will only appear in adult life. Until the age of majority at least, they have the right to innocence. As adults they can then make their own decisions.

For screening to work properly there must be a high standard of medical care and genetic advice, which cannot always be assumed. High standards of care depend on education in the caring professions, but also on a high level of understanding among the public. The more questions the public ask, and the more honestly they are dealt with by health professionals, the better.

The problems of screening seem at first sight less pressing when the screen is for a more benign disease than cancer or Huntington's. As discussed in an earlier chapter, thalassaemia and sickle cell anaemia are genetic diseases that affect the haemoglobin genes. Thalassaemia is very common in the Mediterranean, as is sickle cell disease in central Africa. Thalassaemics are unable to make enough haemoglobin for their red blood cells and, in the severe forms of the disease, require frequent transfusions to keep their red-cell count near normal. Children with sickle cell disease are less anaemic, but, when the amount of oxygen in the blood drops below a critical level, they can develop excruciatingly painful 'sickling crises'.

These two diseases are common because carriers of the mutant genes are protected against malaria. Malaria is now much less of a problem in the Mediterranean, and people from central Africa have been spread to regions of the globe where there is no malaria at all. The anaemia genes are therefore now only remarkable for the diseases they cause. Public awareness programmes and screening for thalassaemia have been carried out in the southern Mediterranean for two decades and the result is already obvious. Children with the disease are no longer being born. A similar outcome will follow for sickle cell anaemia. The defects causing these two hereditary anaemias were among the earliest discovered, preceding the new genetics. On the basis of experience with screening for these conditions, screening is likely to be carried out for other genetic disorders, for it shows that the public can be convinced of the value of genetic screening, even when the genes are common in the population.

Screening works at different levels. Adults who carry genes which may predispose to disease may choose not to have children. If they are carriers of a recessive illness, they may choose to marry only a spouse with whom they can have normal children. Alternatively carriers may wish to submit to antenatal testing in the early stages of pregnancy, so that a foetus with abnormal genes may then be aborted.

Antenatal testing with abortion is very common for the hereditary anaemias and is commonplace for other severe congenital disorders such as Down's syndrome. From a pragmatic point of view, I would suggest most women and their spouses are pleased that the possibility of preventing the birth of children with these disorders exists.

Sex is not a disorder, but it is due to a single gene. Parents very often, and quite wrongly, have strong preferences for the sex of their children. Even in the emancipated and educated West, those who prefer a son to a daughter are in a majority over those who prefer the opposite, and infanticide of female children is common in many uneducated parts of the world. Antenatal testing for sex is not acceptable in our society, but sorting male from female sperm is already almost accepted. The procedure is difficult, and expensive, and undignified, which will mercifully limit its general use. However, one could quite easily imagine a 'do-it-yourself' sperm treatment kit, which would give the same choice to parents, albeit with less accuracy. This horrifies me, although I cannot say why. Perhaps it is the ease with which it could be done.

The technology of manipulating embryos has already been extensively developed in the area of animal husbandry and breeding and the increasing power of this technology opens new problems that will, at some stage, have to be dealt with head on. Genetic screening is now possible when the embryo is only eight cells old. One cell can be taken from this embryo and tested for genetic mutations. The embryo can then be discarded or implanted into the womb, depending on the results of screening. This may be preferable to an abortion for many women, although technically demanding, and by no means simple for the woman herself. It is also technically possible to separate all the cells of an early embryo and to grow a new embryo from each cell. This may be for the good, as it spares the mother the pain of multiple operations to harvest her ova, but again the ethics are not simple.

The 'correct' use of antenatal screening becomes still less sure as the severity of the genetic illness decreases. Adult polycystic kidney disease (APKD), for example, causes intractable renal failure. A pregnant mother would, I am sure, think long and hard about bringing a child into the world who is destined to suffer from kidney failure; such failure inevitably results in a difficult existence, as life becomes dominated by dialysis or the not inconsiderable dangers and disappointments of a renal transplant. However, for those born with the APKD gene, the deterioration of their kidneys does not become a problem until they are in middle age. The treatment of renal failure is improving all the time, and it is quite possible that the problem of organ rejection following transplant may be solved in the next twenty years. Under these circumstances is abortion justified?

There is no right or wrong answer to this question; rather, in a society where abortion is commonplace for social reasons, people will choose for themselves if they want to transmit a condition to their children. APKD is a dominant genetic disease; there are no carriers in the population. Even if only 5% of mothers who carry foetuses with the APKD gene choose abortion, then the prevalence of the disease in the community will steadily decline to the point of complete disappearance. The same will be true of Huntington's disease, and the hereditary cancer syndromes.

It may be argued quite reasonably that to rid the world of APKD or Huntington's disease can only be a good thing. But what of the very common disease genes? The cystic fibrosis gene is carried by one in forty of us. This probably means that it confers some advantage to those individuals who merely carry the gene, as opposed to those who have inherited two abnormal genes and who are severely disabled. It is likely that such an advantage would take the form of resistance to infection, possibly that of tuberculosis. At the moment mankind is in the ascendant over *mycobacterium tuberculae*, the bacterium which causes the illness. The mycobacterium is, however, fighting back, with strains resistant to multiple antibiotics sharply on the increase. Is it possible that by screening the CF gene out of the population we are leaving ourselves vulnerable to more epidemics of TB? An answer is not known. We can look at other populations, such as the people of Africa, and we can be reassured because the CF gene is very rare, and they do not obviously suffer from its lack. The problem is avoided because CF, like thalassaemia and sickle cell anaemia, is a recessive condition. Antenatal screening will only remove embryos or foetuses which have two copies of the abnormal gene. Unless carriers are also aborted, the abnormal genes will not disappear from the population in the forseeable future. However, not all common genetic diseases are recessive, and the problem recurs when we consider a further stage of screening: this might be for genetic variants which are even more common than CF, the genes which predispose to the common complex genetic diseases.

The most pressing example of these is the ApoE4 gene which, as discussed in 'Complex Diseases' (page 119 ff.), shows a strong association with Alzheimer's disease. It is part of a family of genes, the other members being ApoE2 and ApoE3. People with two copies of the gene have a 90% chance of developing Alzheimer's disease at the age of 75, whereas those who only have copies of ApoE2 or ApoE3 have only a 20% risk of getting the illness. Within months of the association between Alzheimer's disease and ApoE4 becoming public knowledge, people from families with the illness were queuing up to take a test for the ApoE4 gene. Shortly afterwards kits for the test were being advertised in *Science* and *Nature*. A few months later still, a French team reported that ApoE2 was associated with longevity.

These circumstances were therefore contriving to produce the worst possible conditions for genetic screening. That ApoE4 was associated with pre-senile dementia was a scientific finding in its infancy. The real risk of dementia could be much less than the early studies had shown. This is not a failing of the original science, it is just how advances in knowledge work; it takes time and more research before results like these can be interpreted properly. The proteins produced by ApoE genes normally ferry fats in the blood. It is quite possible that the dementia associated with ApoE4 is due to clogging of the small arteries of the brain. It may turn out that ApoE4 dementia is not true Alzheimer's disease, but instead a separate condition with a different outlook and different potential treatments. An arterial cause might fit with the longevity apparently conferred by ApoE2.

All this means that testing for ApoE is wildly premature. At the moment there is no cure for Alzheimer's disease and people who have the ApoE4 gene may spend the next forty years of their life under the completely mistaken appehension that they will develop dementia in their old age. Commercial interests are pushing for the widespread introduction of an ApoE4 test. There are, however, no mechanisms in place that will support people who test positive for ApoE4: there is no-one to clarify the result of the test, because the correct answers are not yet known. How can this unfortunate mess be ameliorated? There is no easy answer, but educating the public to be discerning about genetic testing can only be helpful.

The genes that lead to complex genetic diseases should be thought of as 'susceptibility genes', with emphasis placed upon the fact that they do not inevitably lead to disease. Instead they may help people to avoid things in their environment which may harm them.

Asthma, the disease I study, now affects one child in ten, and seems to be increasing. Asthma in children is due to a condition called atopy, which shows itself as a tendency to allergies. Half of the population may be atopic, so it is not possible to consider it a disease at all: if atopy were any more common, non-atopy would be considered the abnormal condition.

The increase in asthma has mostly taken place over the past thirty or forty years. The rate of this increase cannot be put down to changes in our genetic stock. This would take hundreds rather than tens of years, and would require some enormous evolutionary pressure, the existence of which we are quite unaware of. The rise in the number of cases of asthma can only be explained by one mechanism: there has been an alteration in our environment.

The nature of this environmental factor is not known, but a popular and well-publicized candidate is pollution. Parents' cigarette smoking, for example, definitely worsens the disease, as may diesel fumes. However, pollution in Western society is decreasing rather than increasing. Victorian London with its pea-soup fogs was considerably more polluted than present-day London, and yet asthma was then a much rarer condition.

A more likely cause of the increase in asthma is the rise of the house-dust mite. This beast rejoices in the name of *dermatophagoides*, the 'skin eater'. The mite is ubiquitous, being found everywhere in bedding and carpets. In these and other warm, moist areas it can feed off flakes of human skin, which we all shed continuously. The mite dislikes the cold, and so central heating has led to an immense increase in its ability to prosper. In Japan, traditional houses were wooden-floored. Often they were built with a space under the boards to allow circulation of air. As the Japanese have adopted more Western hous-ing and bedding, the mite levels have risen, and so has the incidence of asthma.

But not all allergy or asthma is due to the house-dust mite. In the Nordic countries the spring is usually short and intense. The birch tree is very common throughout Scandinavia, and during the spring all the birch trees come into flower at the same time. For a few weeks the streets and offices are full of sneezing and wheezing people. An inter-esting consequence of this phenomenon was noticed: children born in the three months around the spring carried a higher risk of birch pollen allergy for the rest of their lives. This means that there is a crit-ical period, shortly after birth, when exposure to 'allergens' can induce a life-long allergic state. Richard Sporik and Thomas Platts-Mills showed that the same thing was true for the house-dust mite. If the population of house-dust mites in a baby's house was above a certain critical level in the first year of life, then the child was much more like-ly to develop asthma later in childhood.

For asthma, therefore, screening may accomplish something differ-ent. If children were identified at birth as genetically predisposed to allergies, then measures could be taken to reduce their exposure to house-dust mites or other possible allergenic substances such as cows' milk. Studies by David Hide, on the Isle of Wight, have already shown that changing factors in the environment of new-born children can alter their susceptibility to asthma and eczema in later life.

The example of asthma again highlights the problem of common disease genes; a factor that may be important in complex genetic dis-eases. If 50% of the population are atopic, it is certain that being atopic confers some advantage to people who have an 'atopy gene'. This advantage probably takes the form of an ability to resist infestation with parasites. Geneticists believe that genes which are carried by 1% or more of the population are there for a reason, that is they give some sort of advantage to the carrier. Any form of genetic screening that would reduce the frequency of these very common genes in the popu-lation might lead to very serious consequences.

Other common genes with beneficial effects may be those that pre-dispose to high blood pressure, or late-onset diabetes, or a high level of fat in the blood. The benefit of high blood pressure is unknown, but diabetes genes and genes altering the level of fat in the blood are like-ly to have evolved to deal with dietary insufficiencies.

It is probable, on pragmatic grounds, that screening programmes of some sort will follow the identification of the common susceptibility genes. Will this result in an undesirable effect on the population's gene pool? This might happen if we chose our marriage partners (or our mating partners) on the basis of the results of their screening tests, or if abortion was deemed desirable on the same grounds. For example, two people might forego marriage if their children were to be doomed to death at twenty from myocardial infarction. This would, however, be the case only very rarely. I cannot see people avoiding marriage because their offspring may have a blood cholesterol level that will be 25% above average, or demanding an abortion because their child will have a 50% chance of hay fever from the age of fifteen to forty, and this is the level of risk conferred by the common susceptibility genes. Love is far too irrational a process to be influenced by these sorts of trifling odds. The most likely result of screening for these susceptibility genes is the prevention of future illness.

A common fear is that the information from surveys for suscepti-bility genes will fall into the wrong hands. The wrong hands may be those of the insurance companies, or the State. There is no doubt that insurance companies will do all that they can to get genetic informa-tion, if it is of use to them. Again the correct manner of dealing with this is not clear. It should not, however, be forgotten that, to a large extent, the collection of genetic information is already standard prac-tice: insurance companies routinely ask if there is a family history of premature death, and require testing for elevations of cholesterol, blood pressure and blood sugar. Similarly, population screening for high blood pressure or high cholesterol is already commonplace and widely accepted in our society.

It is therefore very possible that the recognition that prevention is better than cure will lead to the acceptance of genetic methods for sus-ceptibility testing. However, it may well also be the case that the current direct measurement of blood pressure or blood fat level will give much more useful information to prospective insurers than any genetic test. Variation in a gene is one or more steps further away from disease than the reality of a definite physiological abnormality.

Genetic screening does assume a more sinister aspect when it is in any way imposed on the population. In western Europe or the United States a government edict that everyone should be screened for sickle cell anaemia would probably, but not definitely, be rejected as too authoritarian. A similar edict for cystic fibrosis would more certainly be opposed by the populace. However, more subtle pressure from charitable bodies in either case is unlikely to meet much opposition, and might, in the end, have the same genetic effects. The issue here is probably the freedom of choice of the individual.

The problems raised by screening for genes which cause disease are therefore far from trivial. Before the end of the century, twenty or more important susceptibility genes will have been identified, each bringing its own problems. Common sense, the proper education of scientists and public alike, and debate may find a way through the difficulties. However, if testing for disease genes is one thing, testing for genes for 'desirable' traits such as intelligence or physical beauty is altogether another, and implies much that is pernicious. I will discuss these genes in the last chapter.

CHAPTER 13

PANACEA

Screening for genetic diseases is actually about prevention. But the real thrill of the new genetics is the possibility that discovery of disease genes may lead directly to better treatment, or even cures. The route from the discovery of a gene to the discovery of a treatment is not direct, and depends on how and where the gene acts.

The best possible circumstance is that, in its normal state, the gene makes a protein or a hormone that circulates in the bloodstream. The human gene can be cloned and unlimited amounts of the deficient protein supplied for treatment by injection. The success of companies like Genentech testifies that this can work. Diseases such as juvenile diabetes and the anaemia of kidney failure result in deficiencies that can be corrected by the protein products of cloned genes. However, such deficiencies are only rarely genetic in origin, and most genetic illnesses are not treatable by simple replacement therapy.

Sometimes the gene in question may make a receptor for a circulating hormone or other growth factor. If the illness is due to a major mutation in the receptor gene, then the receptor may be irreparable. In complex diseases such as hypertension or asthma, the receptor gene or the gene for the hormone may not be broken, but rather be subtly different from normal. Receptors and triggers such as these can be influenced by orthodox pharmaceutical methods.

The giant pharmaceutical companies, like Glaxo or Sandoz, store many thousands of chemical compounds known in the jargon of the industry as 'small molecules'. Much of the best research into the basic mechanisms of disease is carried out in the laboratories of these companies. The aim of the research is to define a target for new treatments. Such targets are often points of interaction between proteins, such as hormones and receptors. The targets can be within the machinery of the cell as well as on its surface. When a target is identified, the company will work through their compounds until they find some which affect the action of the receptor by mimicking the shape of the hormone or chemical that usually fits into the receptor. A long process of tinkering with the molecule to make it more specific and safe for human use then follows.

Cimetidine, the anti-ulcer drug, was a spectacular result of a targeted search for treatment. Food in the stomach causes certain cells to release the chemical histamine, molecules of which fit into receptors that trigger acid secretion. Cimetidine blocked the histamine from the receptor, and prevented the stomach from making acid. It was safe, and fantastically successful in treating stomach ulcers; it was also outstandingly successful commercially, because the ulcer often relapsed when the course of treatment finished. Ranitidine was a Cimetidine-like drug which had even fewer side-effects than Cimetidine; invented at the British company Glaxo, it is now the world's most popular drug.

Many genetic advances that lead to new treatments will do so because they define new targets for small molecules. To move from a target to an effective compound takes a minimum of five years, and is much more likely to take ten or more. The finding of a potentially useful compound is then followed by years of clinical trials, conducted under the gimlet eyes of the drug licensing authorities. If all the failed treatments that never come to market are taken into account, the costs of developing a successful new drug can reach £100 million and more. It is important, therefore, to realize that a 'genetic breakthrough' reported with a fanfare in the media will never lead to an instantaneous cure.

Again most unfortunately, genetic diseases susceptible to therapy with small molecules or drugs are the exception rather than the rule. The more common case is exemplified by muscular dystrophy. A normal muscular dystrophy gene makes a protein called dystrophin. Dystrophin anchors the contractile proteins to the walls of the muscle cells and is manufactured out of reach, deep within the muscle fibres themselves. It is a 'structural protein': muscle cells without it are weak, just as a concrete building may be weak without an internal support of steel. Small molecules could never fulfil the structural requirements of dystrophin.

Muscular dystrophy will also fail to respond to simple replacement therapy, because dystrophin cannot enter the muscle cells from the blood stream. The Huntington's disease gene works in the brain, also impenetrable to replacement proteins. The genetic basis of thalassaemia has been understood for many years, but replacement strategies are limited to blood transfusions. Cystic fibrosis, the most common of the single gene disorders, affects many different tissues and makes a big protein which has to be folded within a cell to work properly. This means that it, too, is beyond treatment by simple replacement.

More complicated methods of replacement therapy exist. A few genetic immune deficiency syndromes are treatable by bone marrow transplant. This is effective because many cells that make up the immune system are to be found in the normal bone marrow. Some inherited disorders of metabolism can be treated in the same way, although the donor bone marrow will only partially replace the defective enzyme. A bone marrow transplant is not, however, a straightforward operation.

Injecting a foreign immune system into a recipient's body produces two sets of unwanted effects. First, the recipient's (host) immune system may reject the transplant, undoing any benefit of the procedure. Severe immune deficiency helps, because the host is unable to reject the graft. However, the graft of foreign bone marrow is unfortunately capable of rejecting the recipient, causing 'graft versus host disease' which can be very serious, affecting the skin and many internal organs. Both types of rejection need to be treated with potent anti-immune drugs, which have many side-effects, and can leave the unfortunate patient open to infection. Advances in manipulating the immunology of rejection may mean that transplant rejection will be completely preventable within ten or twenty years; bone marrow transplant will then be widely used, but for the present it does not offer a consistent cure for most genetic diseases.

Other means are therefore desirable to deliver a functioning protein or functioning gene into the tissue which suffers from its lack. Here we enter the well-publicized world of gene therapy.

It would seem a fanciful idea that genes could ever be inserted into anything as complex as a eukaryotic cell (a cell with a nucleus). Introducing genes into bacteria is commonplace, but bacteria are simple, and foreign genes can be easily inserted into plasmids that are separate from the main bacterial chromosome.

However, very surprisingly, foreign genes can simply be pushed into a cell. This can be by injection into the nucleus, or even by shooting small glass balls covered with the gene at the cell. Once inside, the genes are quite often taken up into the normal DNA, and they work – or at least they work after a fashion.

This haphazard introduction of genes is the method behind the making of 'transgenic' animals. The eggs of an animal are injected or 'shot' with DNA, and some of the progeny will maintain the foreign genes as their own. In this way goats can be made to produce human proteins, and mice can be made with a preordained propensity to develop cancers. This is very clever, but not altogether a good thing.

The foreign genes are inserted at random into the genome of the host animal. They will almost invariably be in the wrong place; that is, they will be without the sequences that would normally control their proper function. Consequently they might be constantly turned on, or, more usually, turned on only very weakly. Very often the foreign gene will be inserted as multiple copies on different chromosomes. Occasionally the gene will insert itself into a dangerous place, inter-rupting a tumour suppressor gene, or activating an oncogene.

These transgenic animals are therefore genetic monsters, far removed from a normal animal, and often deformed or disabled by the extra genetic material sprayed into their genome. They remind me horribly of H.G. Wells' story of the vivisectionist, Dr Moreau, or of the Soviet experiments of the 1960s where the head of one dog was graft-ed onto another. The manufacture of such beasts has the appearance of science because it uses the tools of science, but the advance of under-standing which underlies science is not in reality being served. It does not make real sense to try to get a goat or a cow to make human pro-teins in their milk. It is simpler to get cells or yeasts to make the proteins in vats and fermenters. Ethically, it also seems much sounder. A well-known example of a transgenic animal is the 'oncomouse', which is genetically engineered to develop cancer. This poor creature, patented by Harvard University, is born not only to die prematurely, but also to suffer terribly from the cancers it inevitably develops and from the experimental chemotherapy which it is marketed to test. There are other ways of testing chemotherapeutic agents which do not rely on sentient animals, and the moral justification of creating such a creature is extremely questionable.

On the other hand, transgenes can readily be inserted into plants, which do not have the same feelings as mammals. The genes which plant breeders are trying to introduce are those which confer resis-tance to insects, or which increase the yield of a crop. In these cases the benefit of the process is reasonably clear. However, the genes may not be stable in their new hosts, or the insects may rapidly become resis-tant to the new plants. There is also a great fear of unknown consequences if these plants are allowed to breed in the open environment. These fears are often ill-informed and Luddite; however, their forceful presentation will ensure that the introduction of the plants will be along controlled scientific lines.

Although transgenic animals often offer uncertain advantage to science or humanity, not all research which alters the genes of experimental animals is wrong. An example is the 'knockout mouse', which has had a specific single gene disrupted. This is a much better scientific experiment than introducing transgenes at random, because it can answer with great certainty the question 'what does this gene do?'. Knockout mice have been used to dissect logically the multiple interacting genes of the immune system, a task that would otherwise be impossible. Knockout mice can also mimic specific human diseases such as cystic fibrosis and these animals can be used to test for new treatments, again in a highly specific and valid way.

Research into transgenic animals has shown that it is possible to take foreign genes and to make them work in animal cells. Defective genes can therefore, in theory, be replaced by the introduction of their normal counterparts. This is the basis of gene therapy, now a hot topic in genetics, but one which faces enormous difficulties that have not been widely discussed.

In the mind of the public, cystic fibrosis is the disease most associated with gene therapy. Several strategies have already been tried to direct the normal CF gene to cells in the lining of the lung by transporting it via a carrier, in scientific terms called a 'vector'.

In the United States a modified influenza virus has been tried out as a vector. The logic behind this is simple, as the influenza virus is already programmed to enter the lining of the lung, and to insert its own genes there. Early experiments, carried out in 1993, showed that the modified virus did introduce a normal CF gene into the lung and that the gene worked just enough to improve some aspects of lung performance. As a few percent of normal function may be sufficient to correct the CF defect, these results were highly encouraging.

One problem, however, emerged almost immediately. The immune system of the CF patients recognized the virus vector, and attacked it. Anyone who has had a common cold has experienced an immune system attack on a virus infection. The lining of the nose reddens and swells and increases its production of mucus. This sort of inflammation is of more severe consequence in the lungs of children with CF. It was seen after the altered virus was sprayed into the lungs of some children, and in one case a child became quite ill after the treatment. There is now a real concern that this vector should be abandoned.

In England, particularly in the laboratory of Bob Williamson at St Mary's Hospital in Paddington, researchers are attempting to use a different vector. Williamson's team are using 'liposomes', which are similar to fat globules, or to very simple cells. Unlike viruses, they do not induce an inflammatory reaction, but can still deliver the gene to its target. The liposomes are directed to the lungs by suspending them in a mist, which the patient breathes through a mask.

Although preliminary results for liposome vectors are encouraging, many further obstacles remain to be overcome. The lung lining replaces itself every few weeks, so that therapy must be repeated at frequent intervals. Inflammatory reactions to the liposomes may eventually occur, or the therapy may not reach critical parts of the airways, leaving the lungs still greatly disabled. Only time and many experiments will give the answers.

In the present state of knowledge, gene therapy resembles the shotgun rather than the rifle in its action on disease. The replacement gene is inserted randomly into the host genome, just as in transgenic mice. Random introduction means that the gene will be without its usual controlling sequences. Because only a little function from a normal gene is enough to correct the CF abnormality, the lack of fine control is unlikely to be a problem. In diseases like thalassaemia, where the production of the two proteins that make haemoglobin has to be matched exactly, full normal control of the replacement gene is a necessity. This is still completely unattainable.

The 'shotgunning' of genes into cells means that occasionally normal genes will be disrupted, and that sometimes these genes will be tumour suppressor genes. Thus gene therapy might cause cancer. Early attempts at gene therapy of bone marrow diseases or immune deficiencies have suggested that oncogenesis may occur, but it is not yet known if this is a result of the therapy or of the underlying disease. Possibly the potential for cancer is much higher in some tissues than others: the bone marrow is full of cells which are rapidly dividing, and is a fertile ground for cancerous growth. By contrast, in non-smokers at least, lung cancer is relatively rare. Gene therapy in the lungs may consequently carry a lower risk than therapy directed at the bone marrow.

Bone marrow and immune deficiency diseases are nevertheless among the most promising targets for gene therapy. The bone marrow is easily obtained. All the cells in the bone marrow, and consequently all the white and red cells in the blood, come from 'stem cells'. These very primitive cells are magnificently called 'pluripotent', because they can produce any type of red or white cell. They are nevertheless nondescript in appearance, and it is only in the last year or two that it has been possible to separate them from the rest of the marrow.

Once the stem cells have been separated, all attempts at gene therapy can be concentrated on them. After replacement genes have been introduced, careful selection can be made for stem cells with new genes which are only in the right place. These safely repaired cells can then be put back into the marrow. There will be no rejection, because the cells came originally from the patient and thus will not be attacked as foreign.

Cancer itself has been promulgated as a target for gene therapy. If cancer is due to oncogenes, then why not just fix the oncogene and cure the cancer? This simplistic argument has attracted much funding. Regrettably, cancer begins with a genetic mistake in a single cell, which causes that cell to increase abnormally in number. Each stage of oncogenesis occurs with a single mistake. In each case it is the progeny of that single cell which carries the cancer to a new level of malignancy. For gene therapy to be effective, it would have to reach every cancer cell; missing just one cell would be enough to cause the cancer to recur. Within the foreseeable future it will be beyond our skill to devise gene therapy which can reach even a sizeable proportion of cells in a solid tumour.

Muscular dystrophy is another potential target for gene therapy. Many cases of muscular dystrophy are due to new mutations: the gene is so long that it is abnormally prone to break. This means that the best screening programme possible will be unable to prevent the disease. It is an insurmountable task to inject all the muscles of the body, including the heart, with DNA containing a normal gene. But this is a case where a viral vector might work. Viruses have sophisticated mechanisms for recognizing certain types of cell, usually through a receptor that is specific to that cell type. Influenza virus only infects cells of the airways, and the Epstein-Barr virus only infects white blood cells. Only those cells have the right lock for the viral key to force entry. Some viruses preferentially infect muscles, and may be just what is needed to carry a normal DMD gene to the place where it is needed most. Because muscle cells do not replace themselves like the cells of the lung lining, gene therapy may only be necessary once, and immunity and inflammation may not be a problem.

There may be ways of introducing genes into cells without indiscriminately spraying them into unwanted places. One possibility is the use of artificial chromosomes. These may be effective vectors for gene therapy, because it is already known that mammalian cells can replicate normally with extra chromosomes (although the extra genes in these chromosomes can cause difficulties). A chromosome would seem to be a very complex structure, not easily given to genetic engineering, but the reality is much simpler than might be imagined. Only three constituent parts of a chromosome, its beginning, its middle and its end, are needed for it to be faithfully copied into all the daughters of a particular cell. The ends of chromosomes, known as 'telomeres', have a special structure which enables them to maintain the chromosome's full length after repeated divisions. The 'centromere' is the middle of a chromosome, and is necessary for the movement of chromosomes into the right place during cell division.

Artificial chromosomes have already been constructed and grown in bakers' yeast. Yeast artificial chromosomes, known as 'YACs', are very simple, containing little other than two telomeres and a centromere. Nevertheless, large stretches of DNA can be cloned into a YAC. Insertion of the YACs in a yeast cell results in successful and faithful copying of the YAC into all the descendants of the original cell.

Human artificial chromosomes do not yet exist, but William Brown in the Department of Biochemistry in Oxford has succeeded in capturing and cloning a human telomere. The human centromere is very bulky, and has been much harder to clone, but eventual dissection of the components which are essential to its function is inevitable. Once reliable human artificial chromosomes have been constructed, gene therapy will be both safer and more sophisticated. The sophistication will follow because it will be possible to assemble all the complex controls for gene function in the correct order in the artificial chromosome.

We are, therefore, only seeing the beginning of the new treatments that will follow from successful gene hunts. Although there is no doubt that in the end we will be better off, we need to concede that the road to any new therapy will be fraught with twists and reverses.

It is also essential to acknowledge again that the new genetics will result in the decline of the population of genes which predispose to disease. Some of these genes are common, and may carry hidden benefits of which we are not aware. The decline will accelerate as new disease genes are identified. This change in our genetic constitution is not being forced on us, rather it is happening through a process of parental choice which will now be difficult to control or to stop. Moreover, if we have already decided that we do not want disease genes in our population, what are we going to do about the genes which we think might be good for us?

CHAPTER 14

THE NEW EUGENICS?

There is a feeling now abroad that geneticists today have been given the tools to dissect completely the workings of life. Just as it has become possible to locate and sequence the genes for complex genetic diseases, so it will be equally possible to map and analyse every human trait. Already the belief is established that it is possible to map the characteristics we believe to be uniquely human, the genes for our behaviour and our intelligence. How genetic are these traits? Will the discovery of all our genes reduce humanity to the prosaic limitations of 100,000 proteins?

At the end of the Victorian age it was argued that character and intelligence were entirely hereditary. This was the natural view of the Right, confirming that the divisions in society were immutable expressions of inborn differences, and absolving the upper classes of responsibility towards their social inferiors. This view continued to hold sway for a further half century; but by the time I was a student, I was able to read books that claimed human beings began life with characters like a blank page. Upon that page parents would write their imperfect prescription for the personality of the child. This was the psychology of the Left: all men and women were born equal.

Until my own children were born I believed the books to be correct. Now it seems obvious that children are born with much of their character intact. A parent or circumstances can always corrupt a child's nature, but I suspect it is impossible to change the child fundamentally. If you doubt the inheritance of character, and you are of early middle age or beyond, look in the mirror. You will not only see your parents' faces, you will also see their mannerisms. The way you scratch your ear, or sit in a chair when you are tired: is that not just like your father or your mother?

Of course not all of the shared behaviour in a family is due to genes, and as a species we are remarkable for our ability to learn. How much then is due to nature and how much to nurture? After a century of debate, the answers are still uncertain. Sadly, all that can be said with certainty is that the genetics of personality and intelligence have always been coloured by prejudice and bigotry.

Since the time of Darwin we have been forced to concede that it is possible to draw striking parallels between our behaviour and person-ality and the animal world. Darwin made us aware that we are not as different from other beasts as we might like to imagine. Because he took man, a creature made in God's own image, into the company of baboons and chimpanzees, Darwin was vilified by many. It was not until the writings of Konrad Lorenz and Desmond Morris that we became reconciled to our animal heritage, and even began to enjoy it. Having once accepted that we share our genetic endowment with ani-mals, it is a short step to agreeing that our genes cause us to behave like them.

Robert Plomin, of Pennsylvania State University, has tried to shed some scientific light upon the genetics of human behaviour. He points out that behaviour is plainly heritable in animals, and that for thou-sands of years we have been breeding animals to select desirable behavioural traits. The cold hand of Science has bred mice of different characters, just as mice have been bred for a propensity to cancer or immunity to infection. This sounds helpful, until we review the list of behaviours which are known to be genetic in mice. These include 'exploration', 'nest building', 'preference', 'withdrawal', and 'aggres-siveness'. We shall return to the significance of these odd traits shortly. For the time being, we should accept that they were chosen for study because they are measurable.

The classical experiment of crossing lines of mice with high and low scores for a particular trait gives a figure of 50% or less for the genetic influence on each of these behaviours. Psychologists in the last sixty years have extrapolated freely from the highly specialized actions of mice in mazes in order to interpret human emotions and ambitions. However, even if we are only trying to understand the behaviour of the mouse, extrapolations of this sort are very unwise, just as it would be foolish to generalize about human behaviour on the basis of a study of humans shut up in prison cells. The mouse breeding experiments prove only that some limited types of mouse behaviour are pro-grammed genetically.

Crossing experiments can yield an estimate of the number of genes controlling an inherited trait. They show that these types of behaviour are more likely to be controlled by several genes rather than one or two. However, the considerable mathematical limitations of this exercise mean that 'several' could be any number between three and fifty.

To understand anything more specific about ourselves than that we share the general animal proclivity to inherit behaviour, we should look to the scientific study of human beings. The first difficulty is to try and measure character. Freud and Jung were the first to classify human nature and to read their books is to understand what enormous fun can be had by fitting behaviour into esoteric pigeonholes. After them came a generation of psychologists who have attempted to measure human behaviour and intelligence; many of these pioneers seem to have shared unfortunate views on race.

A leader of this generation was Hans Eysenck, who opined that there were three basic dimensions to personality, which he called neuroticism, extraversion and psychoticism. Two other dimensions, agreeableness and conscientiousness, were subsequently added to this list. Together they make up a loftily labelled 'five-factor model'. Does this help? The truth is that it does not probably matter if there are three or five factors. Neuroticism, extraversion and psychoticism are part of a jargon that owes its origins to the psychoanalyst's couch. It can be explained that neuroticism approximates to emotional reactivity, and extraversion to sociability. Nevertheless, even when interpreted, the terms have a specialized use which is impenetrable to most people. As soon as one attempts to fit one's own or one's spouse's suppertime behaviour into Eysenck's structure, it is immediately apparent that it forces hopeless simplifications on our motivations and our actions. Even applied to innocent children, it is inadequate to describe the tangle of aspiration and intrigue seen every day in the school playground.

The best that can be said for Eysenck's labels, and others like them, is that they can be measured, or at least pseudo-measured, with involved questionnaires, and that the measurement is reproducible – that is, it gives the same number for the same person on different occasions. The apparent advantage of numbers is that they can be manipulated statistically as a basis for investigating the inheritance of personality.

More recent models of behaviour have been proposed by Robert Cloninger, at Washington University School of Medicine. Cloninger hopes to find the genes for personality, by using the tools of genetic linkage. He hypothesizes that the inherited components of personality modify the 'activation', 'maintenance' and 'inhibition' of behaviour.

He considers activation to be represented by novelty seeking; behavioural maintenance he calls reward dependence; inhibition of
behaviour he calls harm avoidance. These labels approximate to our
own knowledge of the world, which is helpful in understanding them.
It seems to me, however, that they ignore extremely basic behavioural
drives such as hunger, aggression, and the libido, which themselves
vary greatly between individuals. How creativity, or musicality, or the
perception of beauty, or love could fit into such models is beyond comprehension.

I do not mean to denigrate this research, however; I mean only to
illustrate its difficulties. What may be important is that measurements
of personality allow mathematical models of behaviour to be built, just
as enormously complex models of the weather have been constructed.
Ignoring for the moment the reliability of such models, it is just possible they might provide a basis for hunts for genes for behaviour. I am
glad, nonetheless, that this is someone else's research area and not
mine.

Most of us would maintain that intelligence is the unique gift which
sets us apart from other animals. The more ambitious of us would
claim that it is our intelligence which gives us a semblance of God.
Perhaps it is because intelligence is valued so highly that those who
believe themselves to have it in superabundance have behaved so
badly when they have attempted the scientific study of the apparently
less gifted.

Studying intelligence has the same problems as studying personality: it is very difficult to define exactly what intelligence is. We are all
familiar with the intelligence quotient, or IQ, which has now been
replaced with the phrase 'general cognitive ability', or 'g'. The IQ gives
a number which quantifies the ability to reason in various ways or to
solve mathematical puzzles.

Sir Peter Medawar was an Englishman who won a Nobel prize for
immunology. In his later years he flourished as a philosopher of science. Through many of his writings and his public utterances he
attacked the notion of IQ, and exposed the pernicious thought which
it encouraged. He wrote of IQ,

> We must first consider the illusion embodied in the ambition to
> attach a single-number valuation to complex quantities.

He explained that to describe a person's intellectual abilities with a
number such as the IQ is the equivalent of giving a single number to

describe the soil in a field. A single number cannot tell about the soil particle size, or its moisture content, or its ability to grow a crop. It cannot possibly tell about suitability for different crops, or the same crop in differing weathers or seasons, or the response to fertilizer. To say all of these things about a field requires many numbers. Only if we accept that the measurements we make of personality and intelligence are mere shadows, cast by a much more intricate and colourful apparatus, can we assess the evidence for their inheritance.

The best of what we know about the heritability of human behaviour comes from twins. Identical twins are just that, genetically identical. By an accident of division, a single fertilized ovum gives rise to two foetuses and two children. Non-identical twins bear the same relationship to each other as do brothers and sisters; they share half of their genes. By comparing the similarity between identical and non-identical twins it is possible to make a measure of how much similarity is genetic.

In particular circumstances, twins can make it possible to detect some of the effects of their upbringing. For example, twins share more childhood experiences than other brothers and sisters, and this extra sharing can be observed and measured. However, the best experiment is the fate that has occasionally led identical twins to be separated at birth and reared apart by adoptive parents. Here their upbringing will have been different, particularly as it is affected by the personalities of the parents who have adopted them. There are apocryphal stories of identical twins separated from birth who have grown up to take the same sort of job, married women with the same name, and drunk the same sort of beer. Other twin studies show that alcohol consumption and alcoholism have a genetic component, as does the tendency to divorce. These in general exaggerate the genetic component to behaviour and oversimplify the issues.

Robert Cloninger and his colleagues have measured the personality of twins. They report that genetic and environmental factors are of approximately equal importance in 'neuroticism', 'extraversion', and 'psychoticism', and that the genetic component of these personality dimensions was between 40% and 60%. 'Agreeableness' and 'conscientiousness' were, by contrast, not very genetic at all; all but a fifth of 'conscientiousness' was due to the environment. This is difficult to interpret, but it is possible that their measurements are detecting pieces of behaviour that are, to a greater or lesser extent, either genetic or environmental.

There have been many studies of intelligence in twins, and these have been summarized by Plomin. He reports that data from over thirty twin studies, covering more than 10,000 pairs of twins, shows that 'g' is about 50% genetic. But before taking '50%' at its face value, it is worth remembering the case of Professor Cyril Burt. Burt was an enormously respected educational psychologist. In the 1940s he published a number of studies of intelligence in twins. His very definite conclusion was that more than 80% of intelligence was inherited. Medawar claims that the influence of Burt's thought led to the introduction of the 11-plus exam in England, with the unstated aim of separating at an early age those with genetic ability from those without. In the 1970s Burt's scientific credentials came under severe attack, as it became clear that much of his data was poorly collected or even fabricated. Because he was so sure that he was right, he felt no compulsion to obey the proper laws of scientific experimentation.

Plomin and Cloninger and their contemporary colleagues are careful scientists who are not prejudiced by the class-ridden attitudes of Burt or his generation. Their work shows undeniably that there is a substantial hereditary component to both intelligence and personality. It also shows, just as clearly, that the milieu in which children are raised is every bit as important to their final abilities as their genes. This is in marked contrast to such traits as height, where over 90% of the variability in the population is due to genetic influences. However, in adverse circumstances, such as childhood illnesses, environmental influences can override completely the genetic endowment of height. The same applies, with much more force, to intelligence and other psychological capabilities.

Nevertheless, since genes evidently do contribute to personality and intelligence, the next step towards understanding is to determine the number of genes involved. Mathematical analysis of families and twins can be used to estimate the number of genes influencing a particular trait. Just as in the case of cross-bred mice, this exercise, known as segregation analysis, can detect the presence of one or two major genes. Beyond one or two the number is simply unknown. This type of analysis of personality and IQ scores shows no major genetic effects: all that can be said is that a high IQ or assertive behaviour, although half genetic, is not due to one or two genes.

To guess at the number of genes, and their probable relative contributions, it may be helpful to take the analogy of the genes that control our immunity. The nervous system, which contains the physical apparatus for intelligence and personality, is often compared with the immune system. Both have to react to a constantly changing and threatening environment. Both require a sophisticated memory to function.

Hundreds of genes control the immune system. Many of them are polymorphic, and come in different varieties. The polymorphism is important because variety is critical in the constant battle that all living things wage against their surroundings. Variety leads to flexibility of response. The parts of the immune system that come into the most direct contact with the external world are the most polymorphic. The genes for antibodies and for white cell receptors are designed to produce an almost infinite variety of proteins, so that almost any infection can be recognized as foreign and attacked.

The immune system is also characterized by a great deal of 'redundancy'. That is to say, there is an overlap in function between many parts of the system. Redundancy means that, with a few notable exceptions, failure of a particular component or gene in the immune system does not mean failure of the whole; only when many components have failed is there a significant loss of function. Finally, the immune system is characterized by complexity. There are not one or two mechanisms to deal with infection, but many: perhaps more than fifty.

These three features, redundancy, complexity and polymorphism, are likely to be programmed into the genes that influence behaviour and intelligence. We can therefore surmise that there will be very many genes for behaviour, they will interact in complex systems, and that a polymorphism in any single gene will not have an enormous effect on the function of the whole apparatus of behaviour. That this should be the case has implications when we begin to examine what will happen if we can actually identify genes controlling personality and intelligence.

It is very likely that such genes can be found, but only with an enormous effort. It is probable that thousands of families or more would need to be studied in order to localize a gene which accounted for 5% of a particular personality trait. Organized genome searches have been carried out for diseases of altered immunity, such as childhood diabetes. They have been fraught with great difficulties, and have isolated very few genes, which can account for only a few percent of the overall incidence of the disease.

These formidable difficulties have not prevented attempts to find behaviour genes. In July 1993 a paper in *Science* described the finding of a 'gay gene'. The investigators were led by Dr Dean Hamer at the National Institutes of Health in Bethesda, Maryland. They had begun their research by looking at the families of 76 homosexual men. They found that there were more homosexuals in the immediate family than could be expected as a result of chance. They also found that homosexuality seemed more common on the mother's side of the family than on the father's. These observations made them think homosexuality might be genetic, and that it might be transmitted through the mother's side. Although familial clustering can be explained by something in the environment, such as the influence of domineering women, other studies also seem to suggest that homosexuality does have a genetic component.

Because homosexuality seemed to arise on the mother's side, Hamer and his team decided that it might be X-linked, like haemophilia or muscular dystrophy. They set out to scan the X chromosome for a gene which predisposed to being homosexual. They took 40 definite homosexuals and their definitely homosexual brothers, and looked to see if they had inherited the same X chromosome, or part of the X chromosome, from their mothers. They apparently found that 33 of the 40 brothers shared the same segment of the X chromosome.

There are technical problems with this result. Most of the parents of the homosexual brothers had not been studied, so Hamer had to make assumptions on the probability that the X chromosome was shared. If the mother had two X chromosomes that appeared the same in the part that was supposed to contain the gene, or the father's X chromosome looked the same as one of the mother's, then the assumption would be invalid. This type of analysis is perfectly legitimate, but forty sib-pairs is at the lower limit of the number required to make the analysis work properly. Only five incorrect assumptions, a 12% error, would reduce the sharing from 33 out of 40 to 23 out of 40.

My feeling is that the real odds are against the linkage being correct. This does not mean there are not genes which predispose to being homosexual: the evidence is that there are. Nor does it mean that genes for homosexuality will not be found. It only means that this kind of research is very difficult. At the very least, Hamer's research will end in an enormous amount of activity, and may possibly result in the identification of genes. My prejudice is that it is nevertheless highly unlikely that such gay genes would individually contribute anything more than a few percent to an individual's homosexuality. Time, and more research, will tell.

The media activity that followed Hamer's paper in *Science* was premature, because, as Hamer himself was at pains to point out, the case was far from proven. The intensity and the timbre of the media coverage gave John Maddox, the editor of *Nature*, the opportunity to thunder about 'wilful misunderstanding of genetics'. Maddox ignored the real message: it was the results themselves that had ignited the public's fears about genetics. On one side was the anxiety that the presence of a gay gene meant that homosexuality might be treated like a disease. On the other it was argued that parents need not worry about their genetically heterosexual children being corrupted by gay teachers. The largely unspoken concern was that if so complex a phenomenon as homosexuality could be determined by a single gene, then the rest of human behaviour could also be explained equally simply.

A paper appeared in the *American Journal of Human Genetics* a fortnight before *Science* trumpeted the possibility of a homosexuality gene. This paper was just as important, with implications that were just as serious, but it seemed as if no-one noticed. The paper came from Han Brunner and his co-workers at University Hospital, Nijmegen in the Netherlands. They described a large Dutch family in which about half of the males were pathologically violent and aggressive. They were also intellectually sub-normal. This trait also seemed to be passed to the males through unaffected females, suggesting that it was due to a genetic defect on the X chromosome. Sure enough, the gene was unequivocally located on the X chromosome, and was eventually traced to a mutation in a gene for monoamine oxidase A.

Monoamines are a class of 'neurotransmitters', the chemicals that convey signals between the nerve cells in the brain. The monoamines include adrenaline and dopamine, the chemicals that usually mediate arousal, or the 'fight or flight' responses that are brought on by danger. Monoamine oxidases are enzymes that break down monoamines. The effect on the men with the abnormal gene was to bathe their brains in a constant excess of adrenaline.

These results indicate that a single gene can have a dramatic effect on behaviour. How can we reconcile this with theories of multiple genetic effects on personality? The answer to this is to realize that a gene containing a mutation is effectively broken. A broken leg tells us that bones are necessary for locomotion, but they tell us almost nothing about the other factors that allow us to walk, or to run, or to kick a football. For the great majority of us, the genetic influence on our behaviour does not depend on broken genes, but rather on variants of normal genes. Just as the many shapes of noses are due to subtle polymorphisms in nose genes, so are our personalities due to subtle changes in behaviour genes.

It is not known if manic depressive illness or schizophrenia are due to single genes, or if they are the result of the coincident effects of several genes that produce behaviour at the extremes of the normal range. If single genes are eventually shown to cause these diseases, it is likely that they too will be found to be broken. When these genes are found their normal variants will be seen to influence behaviour, but almost certainly not in a overwhelming way. The distinction between variants of normal genes and mutated genes is just as important when we consider intelligence. There are at least a hundred genes that, when mutated, are known to result in sub-normal intelligence; they are of course necessary to the normal function of the brain, but no single one of them can, in another form, confer high intelligence.

The real point of the discrimination between a normal variant and a mutation comes when we acknowledge what we fear. If we can find genes for behaviour and for intelligence, then we might feel tempted to use that knowledge for eugenic purposes.

To understand what is behind 'scientific' eugenics, it is worthwhile learning something of its inventor, Francis Galton. Galton was an Englishman whose life spanned the reigns of Victoria and Edward VII. He had a remarkable curiosity about many things, and it is impossible not to admire his intellectual energy and many of his achievements. Nevertheless, the attitudes to heredity which he so effectively represented hang like a dark cloud over the first half of the present century. His career gives us an insight into how a scientist's observations can be coloured by the society into which he has been born. It also teaches us the harm that is done when a scientist does not try to rise above his limitations and his prejudices.

Galton was born in 1822, the ninth child, after a gap of six years, of a Quaker banker. He was spoiled as a child, his upbringing largely left to the care of a doting elder sister. A fragment of his childhood survives in a letter:

My dear Adele,

I am four years old and can read any English book. I can say all the Latin substantives and adjectives and active verbs besides 52 lines of Latin poetry. I can cast up any sum and in addition multiply by 2,3,4,5,6,7,8, ,10.

I can also say the pence table, I read French a little and I know the Clock.   Francis Galton
February-15-1827

On the basis of this letter alone, an American psychologist called Terman has calculated that Galton's IQ must have been 200, and that he was therefore a genius. He was, however, only a day from his fifth birthday, and the abilities described in the letter are far from unusual in children of that age who have been heavily tutored. It is also relevant to note that Terman was a turn-of-the-century Californian, who himself had virulent views on race and intellect. Even in American society at that time, there was widespread sympathy with the view that the poor, the mentally defective and the criminal were so as a result of heredity. In California in 1918, as L.J. Kamlin has written in *The Science and Politics of IQ,*

> The meek might inherit the kingdom of Heaven, but, if the views of the mental testers predominated, the orphans and tramps and paupers were to inherit no part of California. The California law of 1918 provided that compulsory sterilizations must be approved by a board including 'a clinical psychologist holding the degree of PhD'. This was eloquent testimony to Professor Terman's influence in his home state.

Terman's exultation of Galton's intellect fails to explain how Galton failed to complete his medical studies, and struggled at Cambridge when attempting a degree in mathematics. Galton escaped his unfortunate educational experiences to explore in Africa. On the strength of these experiences, he wrote a book *The Art of Travel; or, Shifts and Contrivances Available in Wild Countries.* This contained some wisdom:

> The traveller who is sick, away from help, may console himself with the proverb that 'though there is a great difference between a good physician and a bad one, there is very little between a good one or none at all.'

And much nonsense:

> Lead [the horse] along a steep bank, and push him sideways, suddenly into the water: having fairly started him, jump in yourself, seize his tail, and let him tow you across.

or,

> When you have occasion to tuck up your shirt sleeves, recollect that the way of doing it is, not to begin by turning the cuffs inside out, but outside in...

The book ran to eight editions. I assume its popularity lay in its capacity for entertaining those who had no intention of leaving England's shores. I hate to think of some red-coated subaltern sifting through its pages for the right advice as the rampant Zulu impis raged outside the stockade.

I suspect that Galton would have left no great stain on the pages of history, except for the work that he began after the age of forty. He pursued an admirable spread of intellectual pastimes. He was an amateur inventor of some distinction, although most of his inventions existed only on paper. As an extension of his geographical interests, he began a systematic measurement of the weather. This most English of pastimes stimulated him in the use and invention of statistical techniques.

His biographer, D.W. Forrest, suggests that it was the infertility of Galton's marriage that prompted his interest in heredity. I maintain that the interest was more attributable to his character. He was a spoilt child from a wealthy family, not used to the company of children of his own age, and not able easily to form friendships. Thereafter, he was able to protect himself from a world obtuse to his sensitivity only by his veneer of intellectual superiority. From this point of view it is only natural he should seek to justify his isolation from most of humanity on the grounds of his own and his class's superiority.

> It is in the most unqualified manner that I object to pretensions of natural equality. The experiences of the nursery, the school, the university, and of professional careers, are a chain of proofs to the contrary.

Galton was not afraid of controversy. Writing a mischievous paper on prayer, he observed the scriptures to express the view that blessings, both spiritual and temporal, should be asked for in prayer. He argued that in examining the effect of prayer one should be guided by averages, and not by specific instances. This, when applied to more temporal matters, is a very sound scientific principle. Galton showed that there was no evidence that prayerful people did better than people who did not pray when they were sick. From a biographical directory he found that the most eminent clergymen, lawyers and doctors died at similar average ages (66.42, 66.51 and 67.04 years). He pointed out that it was usual to offer prayers for the wellbeing of the Royal Family, yet the average age of death of members of Royal Houses was 64.04 years. Missionaries, though acting in the service of

God, often died notoriously young. The editor of the *Fortnightly Review* wrote,

> Your paper is too horribly conclusive and offensive not to raise a hornets' nest.

Galton would have been well pleased at the controversy that followed. It was, however, only light relief from his major work. Three or four years before this attack on prayer, Galton had begun a study of men of high ability, the results of which he published in a book called *Hereditary Genius.*

As a starting point he took *The Times* obituaries of the year 1868. He reasoned that the individuals in these obituaries represented the best of the British race. He found that there were 250 *Times* obituaries per million population over middle age, thus arriving at a figure of one in four thousand for exceptional talent. He then plotted the degree of eminence in various of their kin. It is far from surprising that the families of the great and good contained other high achievers: the ruling classes were, after all, the ruling classes. To this data Galton applied some mathematical analyses. These were very simple, and allowed him to fit the data to his prejudice, but, in his use of averages to compare different groups of people, he had invented the base of modern statistics. His conclusions ran along the lines:

> There is a continuity of mental ability reaching from one knows what height, and descending to one can hardly say what depth. I propose... to range men according to their natural abilities.

He dismissed poverty, nutrition, and education out of hand:

> People are too apt to complain of their imperfect education.... But if their power of learning is materially diminished by the time they have discovered their want of knowledge, it is very probable that their abilities are not of a very high description...

He followed this original work with a study of 286 judges and their families, reaching similar conclusions. When he spoke of his fears about the breeding habits of the lower classes, the agenda of eugenics began to emerge.

> It may seem monstrous that the weak should be crowded out by the

strong, but it is still more monstrous that the races best suited to play their part on the stage of life be crowded out by the incompetent, the ailing, and the desponding.

Galton was far from alone in holding such views. Another Victorian, W.R. Greg, could write in *Fraser's Magazine*:

> The careless, squalid, unaspiring Irishman multiplies like rabbits: the frugal, foreseeing, self-respecting, ambitious Scot, stern in his morality, spiritual in his faith, sagacious and disciplined in his intelligence, passes his best years in struggle and in celibacy, marries late, and leaves few behind him. Given a land originally peopled by a thousand Saxons and a thousand Celts – and in a dozen generations five-sixths of the population would be Celts, but five-sixths of the property, of the power, of the intellect, would belong to the one-sixth of the Saxons that remained. In the eternal 'struggle for existence', it would be the inferior and *less* favoured race that had prevailed – and prevailed by virtue not of its good qualities but of its faults.

Although Darwin, the most eminent of Victorian English scientists, was to play an important part in overthrowing the Victorians' idea of the natural order of things, his own theory of evolution was influenced by the prevalent English class attitudes and disdain for other races. As John C. Greene wrote in his book *Science, Ideology and the World View*,

> It is a curious fact that all, or nearly all of the men who propounded some idea of natural selection in the first half of the nineteenth century were British. Given the character of international science, it seems odd that nature should divulge one of her profoundest secrets only to inhabitants of Great Britain. Yet she did. The fact seems explicable only by assuming that British political economy, based on the idea of survival of the fittest in the marketplace, and the British competitive ethos generally predisposed Britons to think in terms of competitive struggle in theorizing about plants and animals as well as man.

Although Darwin's thought was shaped by these prejudices, he was far too good an observer to be taken in by them completely. Galton was Darwin's first cousin. When *Hereditary Genius* was published, Darwin wrote to Galton, deferring to the work, but also giving his point of view.

I have always maintained that, excepting fools, men did not differ much in intellect, only in zeal and hard work; and I still think [this] an eminently important difference.

In 1871 Galton wrote a paper on 'Gregariousness in cattle and in men'. He took the observation that lead oxen, those which other cattle are prepared to follow in harness, are rare compared to less dominant oxen. He put the proportion of leaders as about one in fifty, and suggested that this was what was required to keep the herd at its optimal size. He then suggested that servility and leadership had a similar occurrence in humans. This is an interesting observation, but Galton then begins the argument that servility should be bred out of a nation and, within a year, he was putting forward the eugenic doctrine in an explicit form:

It may become to be avowed as a paramount duty, to anticipate the slow and stubborn process of natural selection, by endeavouring to breed out feeble constitutions, and petty and ignoble instincts, and to breed in those which are vigourous and noble and social.

Galton's schemes for selective breeding are no more practicable than his methods of crossing a torrent by hanging on to a horse's tail, but the consequences of this sort of thought are found in concentration camps rather than in damp clothing.

Galton continued to express his curiosity about many things, including fingerprinting, word-association as a means of self-understanding, and the study of twins. In 1888 he invented the statistical tool of correlation, the measure of how two or more measurements, such as height and weight, relate to each other. He used correlation as a measure of kinship, foreshadowing the work of Newton Morton by seventy years. Through all of this mental activity he continued to propose the notion that,

The improvement of our stock seems to me to be one of the highest objects we can reasonably attempt.

With this express aim, Galton formed the Eugenics Society in London at the turn of the century. It is reassuring that the Society was never really taken seriously by the public and we cannot blame Galton for originating the pernicious views on race and class with which his name is associated; they were widespread in Europe, and in a different form were also prevalent in the United States. We can, however, blame him for giving them a semblance of scientific credibility.

Galton had a disciple, Karl Pearson, a brilliant man who took Galton's rudimentary methods of comparison and correlation and fashioned them into rigorous statistical techniques that are still in widespread use today. Pearson shared many of Galton's prejudices, as did people who used his statistics in anthropometry and psychology. It would be salutary, as we begin the new era of genetics, for us to remember that these sciences originated in such murky waters.

Eugenic attitudes, and their rebuttal, are not altered by the new genetics. Recognizing fifty genes for high intelligence does not make it easy to select in some arcane and scientific way for the presence of such genes in children. Let us assume, for the moment, that such a selection will become possible. In fifty years it is just conceivable that fifty polymorphic genes affecting intelligence or character will have been identified, and that it will be possible to screen for these genes. The eugenics movement and its apotheosis in Nazi Germany have made it unlikely that any government of a technologically advanced Western society would wish to apply such pressures to its populations. This premise may change at some time in the future, but this is not something that lies in the control of scientists. For the present it is more likely that individuals, rather than states, would wish to apply such knowledge to ensure that their children are of the highest ability. Already, in California, earnest mothers pay highly for a vial of a Nobel laureate's semen, a package of genes for intelligence; however, this type of selection can never compensate for a failure of intelligence on the maternal side.

The alternative might be to examine an embryo for the fifty desirable variants of the genes for intelligence. This would be a far from trivial task. Firstly, both parents would have to be tested, to make sure that they were of the right genetic stuff. An ovum would then have to be brought to conception, probably in a test tube rather than in the normal way. The harvesting of ova from women is already carried out routinely in fertility clinics. It involves hormone injections, admission to hospital, and insertion of several tubes, each a centimetre in diameter, through the wall of the abdomen under an anaesthetic. It is usual for the procedure to be repeated more than once. Women can only find the courage to go through all of this because they desperately want to bear and love their own children.

After fertilization and a day or so of growth in the test tube, a single cell could be taken from the embryo. The tests for the right genes could be carried out on that cell. The embryo could be discarded if genetically inadequate, or implanted into the womb if acceptable.

The above describes the most passive possibility for embryo screening. It is likely that it will be used for screening for genetic illnesses, because it is marginally less traumatic than testing a foetus and undergoing abortion. However, if complex traits like intelligence were wanted, waiting for the chance coincidence of desirable genes in an embryo would be inefficient. A more direct approach would be the genetic engineering of the fertilized ovum. The intention might be to carry out a sort of gene therapy, to replace indifferent genes with their better counterparts.

To do this for one gene is just possible. Because the replacement of genes in gene therapy is random, it requires thousands of cells to be treated, and each cell tested until one is found with the gene inserted into the right place. An embryo is at the beginning of life. This means that almost every one of its 100,000 genes would be called into action as it grew into a baby and then into an adult. We have already seen that gene therapy commonly introduces multiple copies of a gene into the genome. In the embryo, unrecognized insertions of the gene in the wrong place would inevitably take a heavy toll of abnormal foetuses and introduce congenital diseases.

To replace two genes would require a million ova, many times more than a woman is capable of making. Embryo splitting might increase the number of cells with which to experiment, but a million embryos are still probably beyond the limits of the most determined eugenicist. For three genes to be replaced would require a billion embryos. Mercifully, this will never be feasible.

Thus, purely on pragmatic grounds, the widespread genetic manipulation of ova or embryos is highly implausible. The ethical implications of this kind of treatment are almost too enormous to discuss, but one issue is already clear. It is one thing to repair a gene in a diseased individual: when he or she eventually dies the artificial gene will be lost. To replace a gene in an embryo means that the artificial gene will be in all the cells, including the sperm or ova, of the adult who grows from the embryo. The artificial gene is then said to be in his or her 'germ-line'. Once in the germ-line the artificial gene is free to carry its effects down the generations. For this reason alone there is agreement among geneticists that experimentation on embryonic genes should be proscribed.

Bizarre manipulations such as these are therefore already forbidden. More reassuringly, their use to engineer or select desirable traits are almost certainly unworkable. There are more reliable ways of choosing genes for intelligence or an obsessive and over-achieving personality for your offspring. Rather than picking a designer ovum from the back shelf in a dubious shop, it will be far more dependable to find a potential mate with the abilities you seek, and persuade them to father or mother your children.

The great majority of us have chosen our mates in this way: we breed with people in whom we recognize certain characteristics which we find desirable. This method has been popular since the dawn of humanity, and the history of humanity is that the method has always resisted attempts to control it: love, after all, conquers all barriers. Knowledge of genetics is unlikely to alter on any significant scale this fundamental and entirely just way of choosing the genes which will partner our own into the next generation.

The eugenic argument also proposes that genes, or genetic variants, for undesirable traits should be bred out of the population. A great worry about this line of thinking is that those genes are there for a reason, as is any genetic polymorphism that is carried by more than 1% of us. Only recently have we begun to appreciate the importance of genetic diversity in the myriad forms of life on our planet. In the same fashion, it would be most unwise to restrict our own genetic variety. Eugenicists have never had the slightest fear about their ability to choose the right genetic traits to foster. A Galtonesque disdain for people who can make things work has, for example, been common in British educators for most of this century. The results are clear to all. If the eugenicists had their way we would be left with a monoculture of prim, precise individuals with neat beards and immaculately knotted ties, competing bloodily for membership of Mensa while industry ceased to function and crops rotted in the fields.

To understand fully the futility of eugenics, we finally need to come to the remaining half of our characters, that which is bestowed by the environment. The environment can influence many things in many ways. In Galton's time the lower classes were recognized to be much shorter than the members of the gentry or the aristocracy. This surely, it was reasoned, was due to an inferior genetic stock. In fact it was the inevitable result of inferior nutrition. With more universal standards of nutrition the working class and the aristocracy are now of the same height. The present difference is that selective breeding sometimes seems to have left the aristocracy rather short of genes for intelligence.

Society values individuals who have exceptional ability. There have been many scientific or pseudo-scientific studies of exceptionally gifted people. These always allow a genetic component to giftedness, but also frequently identify the childhood environment as critical to later success in life. Great achievers tend either to have had exceptionally fulfilling childhoods, or to have suffered in a way that forced them to develop their abilities. High ability is consistently accompanied by the drive and motivation to practice and improve skills. That this drive is environmental can be shown by studies which demonstrate that high achievers tend to be first in the birth order. The early American astronauts, for example, were all only children or first-born sons. They were chosen for their ability to handle the unexpected and the emergency. This outstanding ability was, it seems, the product of the happy chance that made its possessors the sole focus of parental attention at a critical time of life. Equanimity and self-assurance are therefore unlikely to be genetic traits.

The advent of the new genetics will not produce a new racism or a new wave of prejudice against those perceived to be genetically disadvantaged. Indeed, in many instances it will show such prejudice to be irrational. Attempts to modify the genetic quality of humanity will be no more successful now than they have been in the past. While accepting the existence of genetic differences between people, the way to get the best out of the genes is not to conduct dangerous and dim-witted eugenic experiments. It is to organize our society so that every individual has the chance to develop his or her innate abilities to their maximum.

Because genes for intelligence and character do exist, albeit as members of as yet unknown complex systems, we cannot now avoid their discovery. Even if each genetic polymorphism contributes a small part of the whole, the residual fear engendered by the coming flood of genetic knowledge is that we will be reduced entirely to the sum of our genes; that we will lose all our moral sense once we become aware that all our actions are determined by base pairs of DNA.

It is at this point we can at last set science aside. Although much of what we are is written in our genetic code, history and the great body of literature and the arts concern the ability of the human spirit to rise above limitations imposed upon it. The faculties that make us human are not to be found in mixtures of genetic polymorphism, but in our ability to learn, and to teach. In his essay 'Popper's Third World', Peter Medawar wrote of 'exogenetic heredity'. He invented this term to describe the transmission of knowledge and culture which takes part outside of genes. The ability to make shoes is not inherited; it is taught by one generation to another. Each generation may improve the design of the shoe, or invent a bicycle to do the same thing. Thus a non-genetic evolution has arisen. This evolution takes place in our institutions, so that as civilization advances we see law, justice and democracy arise. Exogenetic heredity is under our control, and we have to accept the burden of deciding how it advances.

What genetics will finally bring is the end of the process of demystification of human beings that began with Darwin. We are only a few steps from the end of the journey that has led us from being creatures made in God's image into the company of monkeys, and now into the undignified simplicity of a three-letter code. The humility which should follow will do us no harm. At journey's end we will be able to prevent or postpone much disease, and we may better understand what we are, but we will still bear the heavy responsibility for the moral use of our genetic gifts.

GLOSSARY

Amino acids   Small molecules which are the basic building blocks for proteins. Approximately twenty amino acids are used.

Autorad   (autoradiograph) X-ray film used to show up fragments of DNA which have been radioactively labelled.

Bases   The building blocks of DNA, known as thymine, adenine, cytosine and guanine, or T, A, C and G. Also known as nucleotides, because of their origin in the cell nucleus. **Base pairs** are matching bases in the two strands of DNA. T always matches with A and G with C. The length of a segment of DNA is often measured in bases, or synonymously in base pairs.

Centromere   The central structure of a chromosome. This is necessary for copying the chromosome during cell division.

Chromosome   A bundle of genes. In bacteria there is only one chromosome, containing almost all the genes. In cells with a nucleus there are often several chromosomes, usually in matching pairs. Humans have 23 chromosome pairs, including the X and Y sex chromosomes.

Cloning   The insertion of a gene or genetic sequence into a carrier molecule (vector). Vectors are grown within hosts, such as bacteria or yeast, to produce an unlimited supply of the cloned DNA or its protein product. A clone is a particular segment of DNA in a particular vector.

Codon   Three bases of the genetic code, a 'word' that corresponds to a particular amino acid in a protein. There are 64 potential codons, formed by all the possible three-letter combinations (triplets) of C, A, G and T. Not all combinations are used. A **stop codon** codes for the end of a gene; a **start codon** codes for the beginning.

Complimentary sequence   The sequence on the second strand of DNA which matches the first. C on the first strand matches G on the second, A matches T and so on.

Cosmid   An artificial circle of DNA into which foreign genes can be inserted. An example of a vector, which can grow cloned foreign genes in bacteria.

Cytoplasm   The part of the cell which contains the metabolic machinery. Distinct from the nucleus which contains the DNA.

DNA   The primary molecule containing genes. Usually in two strands. Made of a sugar (deoxyribose) backbone, and an internal sequence of nucleotide bases. The bases contain the genetic code.

Eukaryocyte   A cell with a nucleus. The nucleus contains the DNA, keeping it separate from the cytoplasm.

Exon   A sequence of DNA from within a gene that is translated into protein. Exons may be interrupted by introns, which are excised from m-RNA before translation into protein. Intron excision is followed by splicing of exons to make a complete gene sequence (*see also* intron).

Gene   The fundamental unit of inheritance. A sequence of DNA bases which codes for a protein.

Genome   The total of all the genes, chromosomes and other genetic material of a particular organism.

Genotype   The changes in the DNA sequence of a particular gene.

Imprinting   (genomic imprinting) The 'marking' of a gene as it passes through the ovum or sperm to render it inactive during the subsequent life of an individual. Thus, genes which are 'paternally imprinted' are only active when inherited from the mother's side.

Intron   A sequence of DNA from within a gene that is transcribed into m-RNA (*see* RNA), but which is cut out before the RNA is translated into protein (*see also* exon).

Library   A representative sample of DNA clones from a particular source, eg. a human chromosome 11 library, or a rat lung library.

Linkage   The localization of a disease gene or other genetic element to a particular chromosomal region. Linkage is achieved by studying families with markers for individual chromosomal segments.

Lod score A statistic which describes the likelihood of a disease being linked to a particular chromosome. A lod score of 3 is taken as proof of linkage.

Marker A sequence of DNA that is used to identify a particular map-point on a particular chromosome. Probes, VNTRs and microsatellites are examples of markers.

Microsatellite repeat A modern type of chromosomal marker, used to make the genetic map.

Nucleotides A synonym for bases.

Nucleus A region in the centre of a cell, separated from the rest of the cell by its own membrane. The nucleus contains the chromosomes.

Oncogene A gene which is associated with cancer.

Phage A modified virus used for cloning genes.

Phenotype The observed effects of a particular gene.

Plasmid A circular piece of DNA which is found in bacteria, separate from the main bacterial chromosome. Foreign genes can be inserted into plasmids and other vectors, a process which is known as cloning.

Polymorphism A variable genetic trait.

Positional cloning See reverse genetics.

Probe A type of marker for a particular chromosomal segment.

Prokaryocyte A simple cell, such as a bacterium, which does not have a nucleus.

Protein A string of amino acids. Proteins are the fundamental building blocks of organisms. Protein enzymes carry out many of the functions required by living things. Each protein is coded by its own gene.

Restriction Enzyme A bacterial enzyme which cuts DNA at a particular sequence. Used for cloning, and for mapping genes.

Reverse genetics The process of moving from genetic linkage in families between a disease and a broad chromosomal region, to identification of the diseased gene itself within the DNA. Also known as positional cloning.

RFLP   A type of chromosomal marker used to make early genetic maps.

RNA   Another molecule containing genetic information. It has a different sugar backbone (ribose) to DNA. RNA may be single-stranded. It is more flexible than DNA and sometimes acts as an enzyme. **m-RNA** (messenger RNA) conveys the genetic code of a gene from the nucleus to the site of protein assembly. **t-RNA** (transfer RNA) transfers amino acids to r-RNA, which transcribes the genetic code into a protein.

Sequence   The order of bases on the DNA or RNA molecule.

Telomere   The end of a chromosomal arm.

Transposon   A DNA sequence that is able to excise itself and reinsert at new locations in the genome.

Vector   A DNA molecule that can be used to carry foreign genes for cloning. Plasmids, cosmids, phage and yeast artificial chromosomes are all vectors. An **expression vector** is a vector which can 'express' its cloned gene by manufacturing a protein.

INDEX